ARCHAEOLOGY IN ANTARCTICA

Archaeology in Antarctica outlines the history of archaeology in the Antarctic and sub-Antarctic.

This book details for the first time all past archaeological work in Antarctica, relating to both its use for conservation and research purposes, drawing on published, unpublished and oral information. This work has addressed historic and current scientific bases, explorers' huts, whaling stations and sealing shelters. The ongoing and long-term research on the sealing shelters and sites in the South Shetland Islands features prominently. The archaeology enables new perspectives on the impact of global modernity and empire in the Antarctic and challenges established dominant discourses on the 'heroic' nature of human interaction with the continent. The work on sealing sites gives voice to the experiences of the sealer as a subaltern group previously largely overlooked by historical sources.

This book will appeal to students and researchers in archaeology, history and heritage as well as readers interested in the human and historical aspects of Antarctica's past and present.

Andrés Zarankin is currently professor of Archaeology at the Federal University of Minas Gerais (UFMG, Brazil) and director of the Laboratory of Antarctic Studies in Human Sciences (LEACH). He is editor of *Vestigios; Revista Latino-Americana de Arqueologia Historica*. His main research interests include historical archaeology, archaeological theory and Antarctic's archaeology.

Michael Pearson is a former heritage consultant and researcher, who now works as an independent scholar. He has worked extensively in historical and industrial archaeology, heritage planning and World Heritage internationally for over four decades, and has undertaken extensive research in the Antarctic, with many published articles and books. He has been awarded the Order of Australia for his work.

Melisa A. Salerno is a researcher at the Multidisciplinary Institute of History and Human Sciences, National Council of Scientific and Technical Research in Argentina (IMHICIHU-CONICET). She has worked on several case studies in historical archaeology, with special emphasis on nineteenth-century Antarctica. Her research interests include the life of 'invisible' groups, power and identity dynamics and embodied practices and experiences.

ARCHAEOLOGY IN ANTARCTICA

Andrés Zarankin, Michael Pearson and Melisa A. Salerno

Routledge
Taylor & Francis Group
LONDON AND NEW YORK

Designed cover image: Detail of "lighthouse" – Raku fired ceramic 2018 – Marcia Seo (Brazil).

First published 2023
by Routledge
4 Park Square, Milton Park, Abingdon, Oxon OX14 4RN

and by Routledge
605 Third Avenue, New York, NY 10158

Routledge is an imprint of the Taylor & Francis Group, an informa business

British Library Cataloguing-in-Publication Data
A catalogue record for this book is available from the British Library

ISBN: 978-0-367-19238-9 (hbk)
ISBN: 978-0-367-19239-6 (pbk)
ISBN: 978-0-429-20125-7 (ebk)

DOI: 10.4324/9780429201257

Typeset in Bembo
by codeMantra

CONTENTS

FIGURES

TABLES

ACKNOWLEDGEMENTS

Many colleagues have given us inspiration over the years. We thank the following in particular for assistance with information and comments on parts of the text.

Gerusa de Alkmim R. (University of Valencia, Spain); Dag Avango (Luleå University of Technology); Guido Bonino (National Council of Scientific and Technical Research - CONICET, Argentina); Jaco Boshoff (Iziko Museums of South Africa); Robert Burton (South Georgia Heritage Trust); John Cooper (Stellenbosch University, South Africa); María Jimena Cruz (CONICET, Argentina); Tess Egan (Australian Antarctic Division); Al Fastier (NZ Antarctic Heritage Trust); Pedro Paulo Funari (Unicamp, Brazil), Russell Gibb (Geometrica Ltd, New Zealand); Lidia Amor (CONICET), Argentina); Louwens Hacquebord (University of Groningen); Janet Hughes (Australian researcher); Adrian Howkins (University of Bristol, UK); James Hunter (Australian National Maritime Museum); Kevin Kiernan (Tasmania); Estelle Lazer (University of Sydney); Eduardo Llanbias (Argentine Antarctic Institute – IAA, Argentina); Alex Martire (ARISE); Angela McGowan (Australian researcher); Lizzie Meek (NZ Antarctic Heritage Trust); Michael Morrison (Purcell Architects, UK); Victoria Nuviala (University of Buenos Aires - UBA, Argentina); Marc Oliva (University of Barcelona, Spain); Claudio Parica (IAA, Argentina); Marty Passingham (Marty Passingham Carpentry, Tasmania); Karen Pymble (Vaughan Evans Library, Australian National Maritime Museum); Marcela Remesal (IAA, Argentina); Peder Roberts (University of Stavanger, Norway); Jesús Ruiz (University of Oviedo, Spain); Flavia Salani (IAA, Argentina); Maria Ximena Senatore (CONICET, Argentina); Fernanda C. Soares (Federal University of Rio Grande, Brazil); Jody Steele (Port Arthur Management Authority, Tasmania); Rubén Stehberg (formerly of Natural History Museum of Chile); Laëtitia Therond (Terres Australes et Antarctiques Françaises – TAAF); the members of the 'Landscapes in White' research project (Laboratory of Antarctic Studies in Human Sciences, University of Minas Gerais – LEACH, UFMG, Brazil) and of the Multidisciplinary Institute of History and Human Sciences (IMHICIHU-CONICET, Argentina). Many thanks to Romina C. Rigone (UBA, Argentina) and André Cancio (UFMG, Brazil) for their help with the style revision and the preparation of the maps, respectively.

Over the years, much of our research was financially supported by the National Council of Research (CNPQ, Brazil); the Minas Gerais State Agency for Research and Development (FAPEMIG, Brazil); the National Council of Scientific and Technical Research (CONICET, Argentina); the National Agency of Scientific and Technological Promotion (ANPCyT, Argentina); and the Milstein Grant from the Ministry of Science, Technology and Innovation (Argentina). We would like to extend our gratitude to PROANTAR-CNPq, the Brazilian Navy, the Argentine Navy and the Clube Alpino Paulista for their logistical support. Thanks to Leica Geosystems for providing free training for operating 3D scanners and making their software available to the 'Landscapes in White' project.

Thanks to Kangan Gupta, Manas Roy, Katie Wakelin and Matthew Gibbons from Routledge for their help in the publishing process.

Special thanks to our families for their love, patience and support.

INTRODUCTION

Antarctica is often referred to as 'the last great wilderness' or a place outside the Anthropocene, an image seemingly supported by its isolation, icy naturalness, exotic animals and the apparent absence of the scars of industrialisation, other than the impacts of climate change and waste taken by sea currents from distant lands, that are now being increasingly recognised (Bargagli 2005). This image, together with the fact that Antarctica lacks a population born and raised there, and the idea that only people 'chosen' for their exploring or scientific skills can get acquainted with the region (beyond the short encounters currently experienced by tourists), helps strengthen a recurring association between the southern continent and a sense of 'otherness' like very few other places in the world (Pena 2016). On some occasions, these feelings of otherness have led to reflections on the spiritual or religious qualities of the territory, and at other times, they have transformed Antarctica into the imaginary home of fantastic and science-fiction narratives (even representing it as a place inhabited by alien creatures and a gateway to unknown worlds) (Leane 2012).

Nonetheless, many parts of Antarctica cannot be defined as 'pristine wilderness': indeed, some locations bear witness to human presence for over 200 years, usually in pursuit of marine animals for commercial profit, and some modern 'scientific stations' approach the scale of large frontier towns. While studies in natural sciences have long dominated research in Antarctica, social sciences and historical investigations including archaeology have their own history in the region. Particularly though not exclusively over the last decades, these fields of study have increasingly proved relevant to 'denaturalise' the well-settled image of the region that connects it with only one of the terms of the dominant Western thought dichotomy 'nature vs. culture' – where nature is assimilated to a primary and pre-cultural matter (Zarankin and Salerno 2014; Leane and McGee

DOI: 10.4324/9780429201257-1

2020). These research proposals are key to broadening the understanding of the cultural processes that are part of not only the uniqueness of Antarctica but also of its multiple connections with the rest of the world.

Antarctica was officially 'discovered' in the nineteenth century. As the child of nationalistic and imperialistic exploration and commercial exploitation, Antarctica has been the object of political contention between countries for over two centuries. The continuous search for maritime resources and the technological advances that encouraged Antarctic whaling in the early twentieth century, combined with growing nationalist sentiments leading up to the Second World War and often expressed in geographical and scientific exploration, accelerated moves to lay claim to Antarctic territory. After the war, and in a new nuclear age, these national aspirations threatened to lead to further international conflict. With the International Polar Years of 1882–83 and 1932–33 as a background, the International Geophysical Year (IGY) of 1957–58 was a multi-national scientific program instigated by the International Council of Scientific Unions and the World Meteorological Organisation that not only aimed to advance post-war science but also demonstrated the possibility of multi-national cooperation. The project allowed for a research exchange between the East and the West, and it required observable data to be available to all countries as a means to advance their scientific usefulness. Among other things, the IGY opened up a means by which the international tensions surrounding the continent could be quietened, and in 1959, the 12 nations that had been active in Antarctica's science and exploration signed the Antarctic Treaty (Argentina, Australia, Belgium, Chile, France, Japan, New Zealand, Norway, South Africa, the Soviet Union, the United Kingdom and the United States).

The Treaty, which entered into force in 1961, set aside, but did not renunciate, existing territorial claims for the duration of the Treaty, and stopped any new claims being made. All activities in Antarctica had to be for peaceful purposes only, and military bases, manoeuvres or weapons testing were banned. The open science approach promoted during the IGY was to be continued. The Treaty now has 54 signatories, of whom 29 have an active presence in Antarctica and have voting rights as 'consultative parties'. The Antarctic Treaty is a unique system of governance: there is no 'government', decisions being made by consensus and implemented through individual national legislation and action (Hemmings et al. 2017; Dodds 2021). The priority of peaceful pursuit of science, the banning of mineral exploitation and the sense of stewardship over the environment were subsequently reinforced by the agreement of the Protocol on Environmental Protection to the Antarctic Treaty in 1991, known as the 'Environment Protocol' or the 'Madrid Protocol'. The Treaty has been called 'a political system based on the authority of science', where '[t]he member states have used scientific research and the useful knowledge it generates to justify the existing political system' (Howkins 2016: 172).

While the Antarctic Treaty set aside territorial claims, a clear underlying element in the operations of Treaty parties is the maintenance of national interest and its recognition (raising questions about potential claims when the Treaty is reviewed). In this process, the list of 'Historic Sites and Monuments' (HSM) created within the Antarctic Treaty System (ATS) to recognise sites of cultural heritage value has been used by many nations to commemorate their national presence on the continent. Many HSM are monuments to national personalities or activities, rather than 'historic sites' as they would normally be defined internationally (Pearson and Salerno in press). Here, it is worth noting that in the process of guiding nominations for UNESCO's World Heritage List, there is an explicit intention to stress the 'outstanding universal value' of sites beyond nationalistic or particularistic perspectives. As will be seen, the specific context in which the HSM within the ATS are nominated has implications for the recognition, research and protection of archaeological sites in Antarctica (in particular, of those sites that cannot be clearly connected with national histories and that do not raise the interest of specific Treaty Parties for their nomination and protection).

In broad terms, archaeology can be defined as the study of people's lives through the analysis of their material world. Archaeologists currently deal with the analysis of multiple research problems, depending on the theoretical-methodological orientation of their particular research. Just to give an example, some of these research problems might include the analysis of subsistence, mobility and settlement patterns, the analysis of sociocultural practices and experiences, identities, power relationships, etc. The material evidence that archaeologists consider for study might range from the food that people ate, the objects and tools they used to perform different tasks, the landscapes they were acquainted with, the way in which they organised their living and workspaces, and so on. Archaeology has many subfields of research. Some are related to what might be called 'periodisation' – a particular slice of human history. The two of most basic periods involve what is usually termed prehistoric and historical archaeology. Prehistoric archaeology includes the study of sociocultural contexts where no written records were available. This implies that researchers only depend on material evidence to learn about the past. Indigenous groups, whose archaeological heritage is often categorised as 'prehistoric', increasingly object to the term, as it implies that they have no 'history'.

Meanwhile, historical archaeology embraces the study of sociocultural contexts where historical sources are available, giving researchers the opportunity to use both material remains and documents as a source of information (Funari et al. 1999). Other definitions of historical archaeology refer to the analysis of the expansion of Western people over the world, and the development and consolidation of capitalism, colonialism and modern society (Deetz 1977; Leone 1988; Johnson 1996; Orser 1996), narrowing the time-frame of the area. Subsets have developed, such as classical archaeology (covering Greece, Rome and the Middle East), Egyptology, colonial archaeology (relating to areas colonised by imperial powers, usually Western) and a myriad of technology-based time phases, such as bronze age, iron age and industrial archaeology. However, in methodological terms, these definitions equally imply a combined use of documentary and archaeological evidence. The study of archaeology in Antarctica clearly draws on both documentary and physical evidence. Some of the sub-Antarctic islands also include evidence of early Polynesian contact or settlement, for which there is no contemporary documentary record.

What is it about the archaeology of Antarctica that justifies a book on that topic? At first sight, there seems to be no reason why archaeology in Antarctica should be any different from that practised elsewhere. But a deeper investigation of both the logistics and techniques of archaeology there, and of the nature of human interactions with the polar environment revealed by archaeology, shows that the study of the cultural heritage of Antarctica and the sub-Antarctic islands has characteristics that are specific to that region (even though they also give a chance to promote comparative studies and discuss the possible connections of the territory with other geographical contexts). Antarctica and indeed the southern oceans are unusual in terms of world history. Antarctica was the last continent to be reached by humans, and the entire history of cultural interactions with the region has been documented to varying degrees in contemporary supporting documents, including journals, logbooks, narratives, commercial and shipping records, scientific papers, photography, film and now the web (Rosa 2016; Stollmeier 2022). Another distinctive feature is that there was no indigenous population in Antarctica, and no subsequent indigenous settlement so far identified. This is also the case for most sub-Antarctic islands, except those south of New Zealand where Māori are known to have visited in both prehistoric and historical times. The archipelago of Tierra del Fuego, which shared some characteristics with sub-Antarctic islands, was also inhabited by indigenous people before the arrival of Westerners, and their descendants still live there (Anderson and O'Regan 2000; Anderson 2005). From what we know right now, the confirmed time depth for Antarctica's history is just 200 years.

It is worth considering what territories are effectively part of Antarctica and what the geographical scope of the book will be. The Antarctic Treaty covers the area extending south of the 60th parallel of the southern hemisphere. Following this understanding, Antarctic territories comprise the mainland and island groups – including the archipelago of the South Shetlands and the South Orkneys. The 60th parallel is politically considered as the northern limit of the Southern Ocean, although the Antarctic Convergence – where the Antarctic colder waters meet the sub-Antarctic warmer waters – varies in latitude, approximately between the 48th and 61st parallel of the southern hemisphere. The sub-Antarctic islands lying between 46 and 60 degrees of the Equator include different groups of islands. Some of them have a geography characterised by tundra, with or without the presence of permanent snow (such as South Georgia, South Sandwich, Macquarie and Campbell Islands, and the Crozet, Prince Edward, Kerguelen, Heard and Auckland Islands, etc.). Unlike the Antarctic, the sub-Antarctic islands are all considered to be the sovereign territory of a particular country. Considering their extreme latitude, the history of the sub-Antarctic islands and the archaeological work conducted there share some of the features of the Antarctic region defined in a stricter sense. Therefore, sub-Antarctic islands have been included in the discussions elaborated in this book. Some other island groups share some of the characteristics of sub-Antarctic islands, but have grasslands and even trees (such as the Malvinas/Falkland, Tierra del Fuego, Ildefonso and Diego Ramírez Islands). These related islands are not dealt with in this volume.

Because Antarctica's human history has been to varying degrees documented in writing from the start, the accepted versions of that history have been traditionally based on the texts. Yet the primary documentary record reflects the objectives and interests of its creators, usually the captains, owners and expedition leaders, and does not necessarily represent the role and contribution of the more anonymous and sometimes illiterate people who made up the bulk of the parties making the occupation of the frigid continent possible. The documentary record is highly selective in what it records; it is fragmentary, largely nationalistic, and focuses on the activities and status of dominant personalities rather than ordinary people (Maddison 2014). Before the late-twentieth century, the literature stresses the adventurous, the heroic, and the qualities of appointed leaders, regardless of their actual success – it is positivist and often self-serving and self-censored. These biases have, until recent decades, overwhelmingly flavoured the writing of Antarctic history in terms of what is usually called a 'master narrative' (sensu Johnson 1999).

This is why the archaeological record is so valuable in providing evidence independent of documentary sources showing how different people actually occupied, modified and adapted to living in the Antarctic environment. Material remains are left behind by people of all kinds, regardless of their status or position. Therefore, the study of the archaeological record has a democratising potential that enables those written out of earlier accounts to be recognised (Funari 1999). This has opened the way for a more pluralistic writing of Antarctic history where ordinary seamen, sealers, whalers and expedition members are included. Moreover, material remains can be the product of multiple activities, some of which could have gone – intentionally or not – unrecorded by written sources. As a result, the archaeological record might shed light on aspects of human interaction with Antarctica previously ignored or underestimated by documents (especially in the case of the earliest records). However, as will be seen, the very fact that the material record studied by archaeology lacks the nationalistic and heroic tenor of most of the historical sources has meant that archaeological sites are often overlooked or avoided by national programs of Antarctic research, while the conservation of specific sites is also very selective (Pearson and Salerno in press).

Despite its potential to account for 'invisible people and groups' in dominant narratives, and despite a growing interest in the lives of ordinary people visiting the region, Antarctic history and archaeology lack two of the recent focuses of scholarship elsewhere in the world. Firstly, Antarctica has no indigenous population, and hence no layer of past indigenous occupation to study. Much of the history of Antarctica has been accompanied by imperialist intent, without the politically and morally unsavoury need to subjugate native peoples (see Bloom 1993; Howkins 2017). While indigenous peoples from other regions participated in the sealing and whaling industries, their archaeological footprint is within the context of those industries, not of indigenous settlement.

The appropriate use of the word 'colonialism' to describe – among other things – human interactions with a region lacking native population is currently being debated (Peder Roberts pers. comm). Notwithstanding, indigenous presence within sealing and whaling operations could have been intertwined with colonial practices at a world stage that had an impact on Antarctica. The colonial gaze could have made indigenous people particularly invisible, as their inclusion in vessels' crews went often undocumented. A similar fate could have been suffered by other minority groups. The case of people from African descent is worth noting. Although their presence among sealing and whaling crews was often recorded, most of their stories have become equally forgotten (Salerno, Cruz, Rigone and Zarankin 2021). Within this scenario, new research proposals are stressing the importance of shedding light on these peoples' lives to commemorate their role in history and create a wider identification between the public and the southern continent.

Secondly, the human presence in traditional Antarctic history was depicted as an almost entirely male domain, until the 1960s when women scientists began to take part in on-ground research at Antarctic stations and in the late 1970s and the 1980s when families, mainly including women in the role of wives and potential mothers of the children born in the region (Llanos 2019), were taken to establish 'settlements' (see below). Despite this masculine view of history, it is highly likely that women sailed as partners or companions on sealing and whaling voyages, but were not mentioned in crew lists. It is also known that women accompanied their husbands serving on whaling stations in the early part of the twentieth century. For example, María Betsy Rasmussen might have lived at the whaling station that her husband, Adolf Andresen – founder of the Whaling Society of Magallanes in 1906 – operated on Deception Island, on the South Shetlands. Also, C.A. Larsen's wife and seven children lived at Grytviken whaling station on South Georgia for at least a season in 1905–06, and a child was born there in 1913 (Chipman 1986: 48). Many women also lived on other sub-Antarctic islands such as the Auckland, Campbell and Macquarie Islands during the whaling and sealing era (Chipman 1986; Burns 2007: 1093).

The first woman known to have landed on the Antarctic continent was Caroline Mikkelsen, who accompanied her husband on the whaling/exploration vessel *Thorhaven* in 1935, and the first to winter there were Edith Ronne and Jennie Darlington on the Ronne Antarctic Research Expedition to Stonington Island in 1947–48 (Darlington 1957; Norman et al. 1998; Ronne 2004; Burns 2007: 1093). The study of gender issues in Antarctica is very active, highlighting the very clear link that exists between masculinity and male cultural perceptions, and nineteenth- and twentieth-century Antarctic imperialism, exploration, politics and 'heroism' (e.g. Bloom 1993; Collis 2009; Dodds 2009). Evidence of the physical presence of women in archaeological contexts is, however, extremely rare, limited to the finding of an American indigenous women's skull in one 1820s sealing site on Livingston Island (Torres 1992, 1999; Constantinescu and Torres 1995; Stehberg and Lucero 1996).

Antarctica is also unusual in that it is the only continent with no human settlements, in the sense of multi-generational established towns or villages. It has scientific stations, which have summer and over-wintering parties, and two attempts have been made to create 'settlements' by Argentine and Chilean authorities to reinforce territorial claims, with families transported to existing scientific bases and 'villages' created – including Esperanza Base on Trinity Peninsula in the first case, and Las Estrellas Village, on King George Island, in the second one (see Llanos 2019; Foscari 2021b). The whaling stations on South Georgia were large and self-contained and could be regarded as 'company towns' (see Chapter 2), but their purpose was solely to service the whaling operations, being abandoned when whaling ceased.

The book title uses the term 'archaeology' in the singular, when in fact there are numerous approaches and uses for archaeological techniques and research in Antarctica, and a plural 'archaeologies' would perhaps be more appropriate (though also more confusing to the uninitiated!) (see Senatore 2019: 756). As will be seen in Chapters 2 and 3, there have been widely different motivations for, and conclusions arising from, archaeological research in the region, particularly noticeable in the different approaches to conservation and research archaeology.

One way of approaching Antarctic history and archaeology is within the framework of modernity, understanding the latter as a complex process intertwined with expansion, capitalism, nationalism, imperialism and science, among others (Leone 1988; Orser 1996). This is useful in providing the context that puts humans in Antarctica, as well as in understanding some of the strategies used by hunting companies and national governments to achieve their goals in the territory. However, these strategies cannot necessarily explain all of the actions that people effectively carried out on the ground. One of the things that distinguishes archaeological evidence from much of the historical discourses of Antarctica is that material remains might reflect individual or group decisions and actions, taken in specific circumstances in a hostile environment, where the relationship of those actions to some distant directive by owners or government, or to societal constructs in the home country, may be very distant. The individual can be influenced by his or her cultural, social or political environment, but those contexts do not always dictate individual decisions and actions, thus opening the way for transgressions, agency, creativity and the need for survival.

Sealers made decisions that best suited their needs at the specific time and place, not usually derived from any overarching directing concept. Late nineteenth- and early twentieth-century expeditioners surviving in mid-winter may make decisions, such as how to dispose of refuse, that would not be tolerated at home. They implemented paper plans for buildings and equipment without necessarily having the skills or the tools to copy those plans accurately. The wrecked sealers or whalers made decisions about personal and group safety, distribution of scarce resources and developing means of escape, which were taken in a situation of extreme stress and uncertainty not often experienced in normal life. The actions of individuals in similar circumstances are often seen to have common characteristics, which should result in patterns within and between sites that are discernible in the archaeological record. But they are actions that would seldom find their way into the written record of the time.

It is important to remember in considering both conservation-related and research archaeology that buildings, huts, shelters and other structures are material evidence of human activity, and as such are also the subject of archaeological study. The archaeologist does not simply make way for the architect because a structure has four walls and a roof, nor should the architect step away from the research of structures that have deteriorated to residual foundations. In this volume, we will expand on the different perspectives on human interactions with Antarctica as revealed through archaeology, considering not only the study of the structures where people sheltered but also investigating other expressions of material culture necessary for their subsistence and work. Human activities in the sub-Antarctic islands started a little over 250 years ago, and in the Antarctic itself – as previously mentioned – just 200 years ago. This interaction coincided with the industrial revolution, the birth of modern science, the continued growth of sea-borne empires, global exploitation of natural resources and the mass migration of populations around the world. As a result, Antarctic history is infused with global industrialisation, capitalism, imperialism, nationalism, international politics and scientific exploration, and these themes influence the understanding and interpretation of the archaeological evidence.

The book comprises four chapters. In Chapter 1, we outline further the context of human interactions with the Antarctic environment as a background to archaeological work there. Chapter 2 looks at the history and scope of archaeology in the Antarctic and sub-Antarctic regions, particularly that associated with heritage conservation work. Meanwhile, in Chapters 3 and 4, we focus on the research archaeology being undertaken in the South Shetlands Islands. This distribution of description and discussion might appear at first sight uneven, but as we will show, the archaeology aimed at gathering information for site management and conservation planning which has dominated earlier research in the sub-Antarctic and most of Antarctica has been sporadic, and most of the effort has gone into management and building conservation processes rather than the archaeological research. Though not exclusively, much of this work has dealt with the sites used by late nineteenth- and early twentieth-century explorers. In the South Shetlands, however, there has now been a quarter of a century of research-oriented archaeology aimed at expanding our understanding of the sealing industry there, the first human activity on the continent, and placing that enterprise into the global context of the modern world.

1

DEVELOPING AN UNDERSTANDING OF ARCHAEOLOGY IN ANTARCTICA – HUMAN INTERACTIONS WITH A POLAR ENVIRONMENT

1.1 The historical and geographical background

It is easy to forget how much the world has changed since humans first set foot in Antarctica 200 years ago. Antarctica was the only continent first visited and inhabited during the industrial revolution, a phenomenon that gave rise to the modern world for good and bad. In 1775, the year Captain James Cook crossed his 1772 track and completed the first circumnavigation of Antarctica proving its insularity, James Watt and Matthew Boulton began their 25-year partnership that revolutionised the production of the steam engines that drove the industrial revolution. George Stephenson in 1816 patented a steam locomotive that would lead to his *Locomotion No1* that was the first in the world to carry passengers on the Stockton and Darlington Line in England in 1825. These dates straddle the first sealing expeditions to the South Shetland Islands, the first human occupation of Antarctica. The Napoleonic Wars ended just four years before the first sealing expeditions in 1819–20.

An Antarctic zone had been hypothesised by Greek philosophers as far back as the fourth and fifth centuries BC. A southern landmass was thought by some to be necessary as a counterweight to the known lands of the Northern Hemisphere, and in 322 BC, Aristotle named the southern region Antarktikos, in counterpoint to Arktos, the northern region lying under the constellation of the Bear. Both regions were thought to have frigid and entirely uninhabitable zones at each pole of the world (Akkerman 2016). To later enlightenment explorers such as James Cook, who circumnavigated the world in high southern latitudes looking for Terra Australis (an extended landmass in the Southern Hemisphere) and was the first known navigator to cross the Antarctic circle, there had to be a frozen land to account for the fresh-water floating ice of the circumpolar seas that prevented him from moving further South. While Cook did not see it, he knew

DOI: 10.4324/9780429201257-2

it had to be there; in Nicholas Myers words: 'It was the scent of it rather than the thing itself' (Myers 2016: 38). Cook's voyages defined the envelope within which the frozen continent had to exist, and later explorers continued the hunt for the real thing.

The Russian explorer Fabian von Bellingshausen in the *Vostok* emulated and exceeded Cook's circumnavigation in 1819–21 and succeeded in making the first verified sighting of the Antarctic continent (probably the Finibul Iceshelf) on 28 January 1820 (Debenham 1945). Others slowly identified more areas of coastline, some of them sealers and whalers (see Pearson et al. 2020 for more detail). Within the context of modernity, the exploration of Antarctica was driven by three, often intertwined, motives: national rivalry, imperialism and the promise of commercial exploitation. The post-French Revolutionary and Napoleonic Wars saw a revision of European national power relationships, and the economic and political independence of the United States of America in the wake of the American War of Independence and the War of 1812. Britain, France and Russia vied to revive, expand or maintain their empires and to stamp their scientific and exploration credentials on the post-war world map, as the United States widened its commercial interests. At the same time as nations were exercising their global reach, entrepreneurs, individually or representing companies, were extending their commercial exploitation of the sea, contributing to the global expansion of capitalism (Senatore and Zarankin 1999; Zarankin and Senatore 1999, 2005; Maddison 2014; Pearson 2018c; Salerno et al. 2019).

Marine animals were the prey of different cultures over time. Notwithstanding, traditional hunting is different from capitalist exploitation, as the former represents a small-scale activity intending to meet the needs of a specific group, and the latter has a broader scale with the aim to make a profit through commercial interactions, ruled by the logic of supply and offer, and cost and benefit (Stevenson et al. 1997; Gillespie 2005; Motos and Wilson 2006). In the seventeenth century, commercial whaling started developing as a means to get a variety of products which became increasingly valued at an expansive market. Whale oil was used for illumination, and it also became relevant as a machine lubricant and cloth-softening agent in the context of industrialisation. Whalebone and ambergris were used for producing corsets and umbrellas, and fixatives for perfumes, respectively (Verrill 1916; Creighton 1995; Currie 2001). Commercial whaling initially focused on the Northern Hemisphere, and it took place near the coasts. But as the presence of whales diminished in surrounding areas, and as new advances were made in vessel technologies, whalers moved further away reaching the Southern Hemisphere. The exploitation of seals and whales were separate industries. However, in some cases, whalers also relied on sealing to add to their cargo and enhance profitability, and the techniques and instruments used to get some products were often similar.

Commercial sealing gained particular relevance in the late eighteenth century when the population of sea otters seriously diminished, and hunters found out that seal skins could provide an alternative to sea otter skins for the clothing industry (Clark 1887). Sealers were not only interested in skins but also in oil. Because of the distribution and abundance of animal colonies, the South Seas became a focus of sealing operations. Within this vast region, sealers sought the hides and dense underfur of the fur seal, members of the *Arctocephalus* genus that inhabited different regions of the Antarctic, sub-Antarctic and adjacent areas: *A. gazella* (Antarctic fur seal); *A. tropicalis* (sub-Antarctic fur seal); *A. pusillus* (Australian and South African fur seal sub-species); *A. forsteri* (New Zealand fur seal); *A. philippii* (Juan Fernández fur seal) and *A. australis* (South American fur seal). Moreover, sealers looked for the oil of the elephant seal (*Mirounga leonina*). The skins of southern sea lions (*Otaria flavescens*), commonly known in the trade as 'hair seals', were also sought in the islands and coasts such as Patagonia, Tierra del Fuego and the Malvinas/Falkland Islands. A sealing voyage might focus on one or the other, or might take an opportunistic approach and gather whichever type of seal offered the best prospects at the time.

Sealing in the South Seas was mainly carried out by private companies, owned by one or more individuals who provided the capital for a sealing voyage (including the vessels, the equipment and some basic stores). The owners hired workers, who were paid a lay or share of the catch in exchange for their sailing and hunting duties. The actions of companies in exploring previously unknown territories (see below) eventually had an impact on the early incorporation of these regions into the modern world, in contrast to the more normal process in which specific nation-states sought to declare sovereignty over new lands. Nonetheless, it has been suggested that the presence of sealers in recently discovered regions could have served as an advance guard for their countries – contributing to their economy and to their imperial or colonial aspirations, intentionally or not (Mayorga 2020). Even though sealing companies were based in multiple countries, those from Britain and the United States played a prominent role in the industry. The wealth that sealing operations brought with them contributed to the growth of several cities in both countries. The history of the port towns located on the east coast of the United States (where many whaling and sealing companies had their headquarters) bears witness to this and the tight relationship that distant geographical contexts had as part of a world economy.

Seal skins were sold in different markets, initially including Canton (China) and later London and New York, among others. Oil was more widely demanded. But what were these products used for? Fur seal pelts had dense underfur and were used in the same way sea otter and beaver furs were used, either as a pelt with the fur still on the skin or as fur shaved from the skin and felted. The pelt could be used as a top-end clothing decoration, and it was possibly used to some extent this way in China in the late eighteenth century, especially for the neck and cuffs of gowns and to line clothing (Kirker 1970: 8). Seal pelts were also used in the making of caps and other types of hats (Burton 2018: 89). Thomas Chapman, a London trunkmaker, was credited with the development in the 1790s of the process enabling the removal of guard hairs from pelts so the dense underfur could be removed for felting, a technique that was believed to have been originally perfected in China. Chapman indicates that the felted underfur was used for hats, in the same way as the more common felted beaver hat, or spun to make fine cloth that was used in luxury items such as shawls (Chapman 1810: 9; Busch 1985: 8) – while pelisses and spencers, coats and breeches are mentioned elsewhere (Burton 2018: 90). The guard hairs were sold for stuffing cushions. The skins of fur seals after the removal of the fur, and those of the sea lions, could be tanned for leather. Chapman himself bought seal skins to cover trunks to make them waterproof, and skins were used also to make shoes, gloves and book bindings. The skins could also be rendered down to extract size, a glutinous substance used for dressing fabrics and paper (Burton 2018). Elephant oil, which was very similar in quality to good whale oil, was used, like whale oil, as lamp oil, fine lubricating oil for machinery, and for fulling wool prior to spinning, finishing textiles and ropes, tanning and finishing leather, leather dressing, and in the manufacture of soap, paints and varnishes (Busch 1985: 164; Ellis 1991).

The exploitative nature of commercial sealing promoted constant geographical expansion. Although sealers tried to remain silent about the location of the rookeries they found, it was only a matter of time for the secret to leak out. As the competition for resources increased and the size of colonies shrank, sealers started searching for new and more profitable hunting grounds (Busch 1985; Zarankin and Senatore 1999; Caviglia 2015). Sealers operated in remote locations where state control was absent or minimal. In the South Seas, the sub-Antarctic islands were more accessible than Antarctica itself, and attracted attention earlier. Table 1.1 summarises the spread of sealing throughout the southern oceans, and as can be seen, the two Antarctic island groups of the South Shetland and South Orkney Islands occurred late in the process. Figure 1.1 provides a map of Antarctica, the sub-Antarctic islands and some other locations referred to here and in the rest of the book. The most intensely sealed island groups were South Georgia, the South Shetlands, Kerguelen and Heard Island. The archaeological evidence for marine mammal exploitation and scientific research sites on the sub-Antarctic islands is discussed in Chapter 2, while that for the South Shetland Islands is presented in Chapters 3 and 4.

A sealing ground not covered in this book is that on the Tierra del Fuego archipelago, at the southern tip of South America. While this region lies between 53° and 56° south latitude, the equivalent of many of the sub-Antarctic islands discussed here, and its climatic conditions are decidedly sub-Antarctic, its close connections with the American continent and its resident indigenous population set it apart from the areas on which we focus. Nonetheless, it was an important part of the sealing world (Clark 1887; Busch 1985; Mayorga 2020), maintaining multiple relationships with the operations carried out in Antarctica (Salerno, Cruz and Zarankin 2021). As an example, sealers visiting the South Shetlands frequently stopped at different points of Tierra del Fuego to get water, food and wood; make exchanges with indigenous people; repair the vessels at safe and well-known anchorages and obviously to take seals (sometimes completing the cargo caught in Antarctica). Especially at the end of the nineteenth century, some

TABLE 1.1 Expansion of the seal hunt in the Southern Hemisphere

First recorded takes	*Place*
1591	South Africa and Namibia
1766	Malvinas/Falkland Islands
1786	South Georgia Islands
1788	Tierra del Fuego and Patagonia
1788	St. Ambrose and St. Felix Islands
1789	St. Paul and Amsterdam Islands
1790	Tristan da Cunha
1791	Kerguelen
1792	Más Afuera (Alejandro Selkirk, Juan Fernández Islands)
1792	New Zealand
1798	Bass Strait and Australian coast
1799	Prince Edward Islands
1804	Chatham Islands and the Snares
1804	Crozets
1804	Antipodes Islands
1807	Auckland Island
1810	Macquarie Island
1810	Campbell Island
1819	South Shetland Islands
1821	South Orkney Islands
1823	Diego Ramírez
1855	Heard Island

sealing companies were based at Punta Arenas, located on the Strait of Magellan (Payró 1898). Among other things, this port may have been used by some sealers visiting the Antarctic sealing grounds in the latter part of the nineteenth century to dispatch their cargo, renew their crews and buy different kinds of supplies (Salerno, Cruz and Zarankin 2021).

Almost all of the archaeological work in Tierra del Fuego to date has investigated the history of settlement by the Selk'nam (Ona), Kawésqar (Alacalufe), Haush and Yaghan (Yamana) groups and their predecessors, going back to 8,500 BC but becoming far more populous about 2,000 years ago (see Borrero 1999; Esteves et al. 2001; Zangrando 2010; Martin et al. 2015). Subsistence depended on guanaco (in the north), marine mammals and molluscs. Sea lions and fur seals began to be commercially exploited by European sealers from the late eighteenth century, and sealing was intermittent there from that moment on. Indigenous groups sometimes provided sealers with seal skins and some other products in exchange for Western articles. Native people may have been taken aboard sealing vessels – voluntarily or not – for their knowledge of local geography and the distribution of resources, as well as for their hunting abilities (Stehberg 2003; Mayorga 2016, 2020). Sealers' catches could have had a significant impact on the subsistence and diet of some indigenous groups (Schiavini 1990; Orquera and

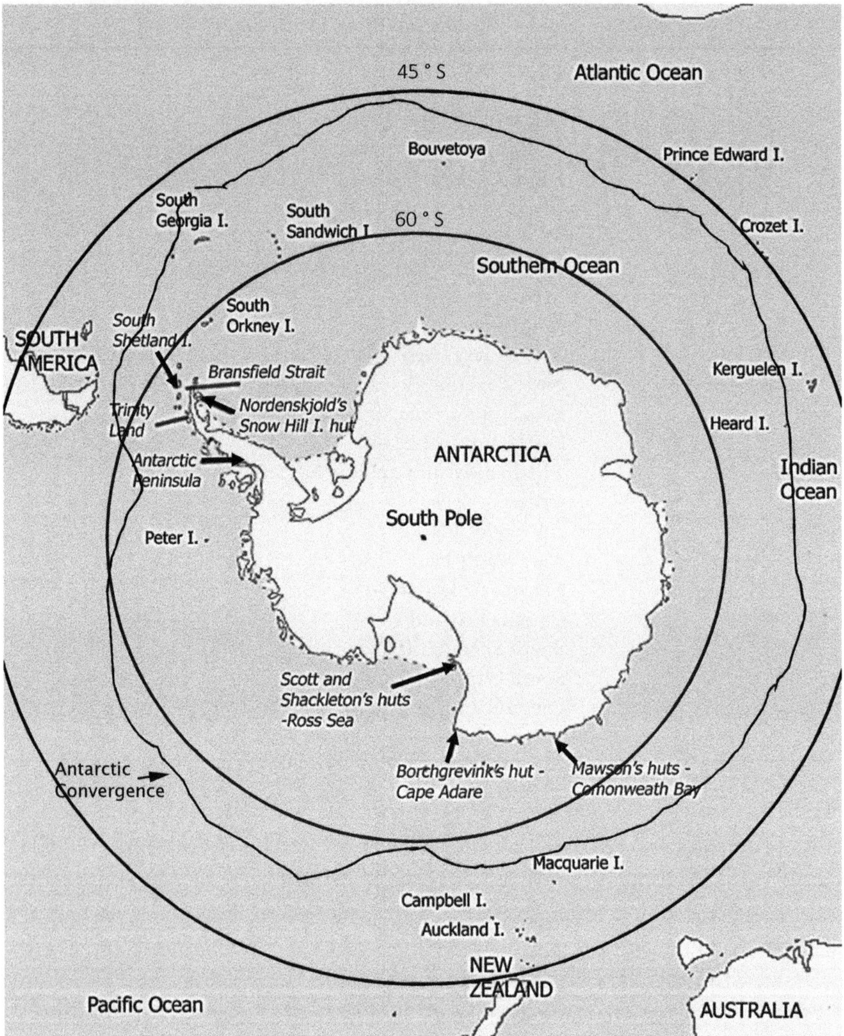

FIGURE 1.1 Map of the Southern Ocean and Antarctica showing the sub-Antarctic islands and Antarctic locations (based on Wikimedia Commons 2004).

Piana 1995). Some historical sources refer to the violence that some sealing crews could have exerted over indigenous groups, including abuses committed against women (Payró 1898). Possibly the sealers and definitely later pastoralists and miners brought with them diseases that were catastrophic for local people. Many were lost to measles in 1876, and the numbers of Yamana dropped from perhaps 3,000 to only 1,000 between 1881 and 1883 due to measles and smallpox (Cook 2004: 65; Chapman 2010: 471).

A project to study early sealing sites historically and archaeologically in Tierra del Fuego commenced in 2018, with initial survey of the eastern tip of the

Isla Grande, and to extend to other parts of the archipelago as resources permit (Salerno et al. 2018; Salerno, Zarankin, Cruz, Rigone and Weissel 2021). Within this context, it is worth mentioning that the presence of archaeological remains left behind by sealing crews has been documented by the Management Plan for the Provincial Reserve of Isla de los Estados (n.d.), Argentina. Notwithstanding, a systematic approach to these remains is still needed. Archaeologists have recently studied the zooarchaeological remains connected with a twentieth-century sealing factory on Bahía Thetis, on the eastern area of Peninsula Mitre, Isla Grande. Although this factory does not represent early operations in the region, as the ones described in the book for the South Shetland Islands, it clearly accounts for another stage in the long history of sealing in Tierra del Fuego (Vázquez and Santiago 2014; Vázquez and Zangrando 2019).

The first humans documented in the literature to have landed on and to have had a meaningful interaction with Antarctica and its close islands were the sealers in the South Shetland Islands, which lie just over 100 km off the northern tip of the Antarctic Peninsula, and about 820 km south of Cape Horn. The widely accepted and clearly documented version of the 'discovery' points out that William Smith, a British sealer and merchant ship master, first sighted the islands on February 19, 1819, when blown south on a passage from Buenos Aires to Valparaíso in the brig *Williams*, and repeated his visit in October of that year, sending a boat ashore to claim the territory for King George III. The Senior officer of the British Royal Navy in Valparaíso then hired the *Williams* to follow up the discovery, placing Edward Bransfield, Master of HMS *Andromache* aboard (with Smith as pilot) under orders to carry out an exploratory voyage (Jones 1985; Campbell 2000: 40–41, 69–70). The islands were observed to have large numbers of fur seals on those beaches that were free of ice. News of Smith's South Shetlands soon reached the ports of Valparaíso and Buenos Aires, and quickly moved on to New England and British ports.

Even before Smith's discovery was officially confirmed, sealing captains based in Buenos Aires seemed to have been in search of the new islands, quickly followed by those from Britain and the United States, as news about the location of the new hunting ground could have rapidly spread when the Williams reached South American ports. Alternative hypotheses on the discovery have even speculated that sealing vessels from Buenos Aires or the United States could have visited the South Shetlands before February 1819, considering the huge amount of skins taken in the period and the possibility that captains had remained silent about the new sealing ground (Fitte 1962; Bertrand 1971). However, these ideas are not supported by verifiable evidence. Some other versions of the discovery do not involve sealers. One of them refers the potential sight of the archipelago by Dirck Gerritsz, commander of the Dutch ship *De Blijde Boodschap*. The other one considers the possible sighting or landing on the archipelago by the crew of the Spanish warship *San Telmo*, which was reported to lose her masts and rudder in the Drake Passage at 62°S on September 4, 1819. However, while in the first case the discovery claim was questioned even by Gerritsz' crew (Howgego 2003), in the second one it has proved impossible – despite extensive research – to confirm if survivors sighted or landed in the region, and if the wreckage recorded on the South Shetlands was from the ship (though this seems likely and is described further in Chapter 2). The claims in the recent Historic Sites and Monuments (HSM) listing of the *San Telmo* within the Antarctic Treaty System (ATS) seem aspirational rather than founded in evidence, and the claimed location of observed remains of this wreck on Livingston Island has been recently convincingly challenged (Cazenave 2019). Whether a shipwreck equates with discovery is a moot point (Aragón 1991; Pinochet de la Barra 1991; Pearson et al. 2020: 263).

As previously mentioned, this intense interest on the South Shetland Islands was the result of the constant need of the now global sealing industry to find new sealing grounds to replace those where the seals had been exterminated by the unregulated and indiscriminate nature of the industry (Busch 1985; Richards 2003). The South Shetland Islands were visited by three sealing vessels in the first hunting season of 1819–20, but some 120 known sealing visits to the islands were made in the 1819–20 to 1821–22 seasons (Figure 1.2). Sixty-five of the latter were from Britain (of which four at least were part-owned or based in Chile or Argentina), 52 from the Unites States and three from Australia (Pearson 2019; numbers based on information collated from Headland 2009 and other

FIGURE 1.2 Hersilia Cove on Rugged Island looking north-east over New Plymouth (President's) Harbour to the Byers Peninsula on Livingston Island. These two harbours were used by the sealers in the 1820s (Photo by Michael Pearson 2005).

sources). A minimum of 300,000 seal skins were taken in that period. Only some years later, however, the seal numbers had been so diminished that the South Shetlands were no longer profitable. Thus were the South Shetlands absorbed into the modern global economy. The archaeology of the over 50 sites mainly resulting from this period of intensive exploitation – and generally consisting of the remains of humble stone structures and associated artefacts for work and subsistence – is outlined in Chapters 3 and 4.

The collapse of seal numbers put an end to the first phase of human interaction with Antarctica. A second phase of sealing extended between the 1830s and the 1850s, and a third one between the 1870s and the 1890s. In each case, the number of vessels visiting the Islands was fewer (30 and 50 American vessels, respectively) and the number of seals taken was much reduced from its former levels, the populations not having fully recovered (Berguño 1993b; Headland 2009). Although researchers tend to focus on the first phase, the later phases also deserve attention, as they are not only part of the history of sealing on the South Shetlands, but they can also provide information on the changing interactions between sealers, the Antarctic environment and animal colonies over time (Salerno and Cruz 2019).

Whaling from sailing ships with open whaleboats in southern waters focused initially on the Southern Right whale (*Eubalaena australis*) which was slow swimming and had large amounts of oil and baleen (whalebone), the Sperm whale (*Physeter microcephalus*) for its high-quality oil and the slow-swimming Humpback whale (*Megaptera novaeangliae*). When factory ships with steam chasers equipped with harpoon guns were introduced in the late nineteenth century, the faster rorqual whales could be hunted as well: the fin whale (*Balaenoptera physalus*), the blue whale (*Balaenoptera musculus*), the sei whale (*Balaenoptera borealis*) and the minke whale (*Balaenoptera bonaerensis*). This transformation of whaling in the South Seas began with the South Shetlands, South Georgia and the Antarctic Peninsula and heralded a new period of exploitation of Antarctic marine resources – one that extends to the present time and has deeply influenced the development of the ethic of environmental protection that now dominates Antarctic politics. Initially, whaling took place from factory ships. Industrial-scale shore-based whaling stations first appeared at Grytviken on South Georgia, in 1904, and on Deception Island in the South Shetlands, in 1912 (Tønnessen and Johnsen 1982; Basberg 2007: 1072–1076). These places saw the first large-scale human settlements established there (Figure 1.3).

While up to this point, we have referred to the sealing and whaling industries in the southern seas, it is worth noting that penguins have been also taken for oil production in different scales on the sub-Antarctic islands. The exploitation of the natural resources of the region has been devoted not only to commercial but also to domestic purposes. Birds, bird eggs, fish, seals, whales and vitamin C-rich Kerguelen cabbage (*Pringlea antiscorbutica*) and Macquarie Island cabbage (*Stilbocarpa polaris*) were used for food on the sub-Antarctic islands and (without the cabbage) in the South Shetland and South Orkney Islands by hunters and explorers. All of these activities have left an archaeological footprint.

FIGURE 1.3 Stromness whaling station, South Georgia, 2007, demonstrates the large scale of the structures at whaling station and the associated conservation and environmental challenges they pose (Photo by Jens Bludau, Creative Commons).

An expression of imperialist aspirations that emerged in the mid-nineteenth century was through the exercise of scientific and geographic 'mastery' of the Antarctic environment. Adrian Howkins (2017) convincingly portrays this in the twentieth century as imperialism by the demonstration of 'environmental authority' (see more in Chapter 2). As a background reference, it is worth noting that soon after the 1820s sealing boom, HMS *Chanticleer*'s 1828–31 voyage under Capt. Henry Foster incorporated scientific study of the South Shetland Islands in its wider scientific investigations. Furthermore, the naval explorations of the 1830s and 1840s (commanded by Jules d'Urville, Charles Wilkes and James Clark Ross) were supported by scientific objectives (Mawer 2006; Pearson et al. 2020). The explosion of expeditions to the mainland from the 1890s was underpinned by the aim of expanding the frontiers of scientific and geographic knowledge (Larson 2011). Within this context, 15 major expeditions took place between 1897 and 1917, with the exception of the Japanese expedition under Norbu Shirase, entirely dominated by European nations or their former colonies (in the case of Australia). Table 1.2 provides a list of these expeditions, as it will be relevant for understanding references to archaeological works provided in Chapter 2.

This period is widely known as the 'heroic era' of Antarctic exploration, a label coined in the 1950s in an attempt to highlight the challenges that exploring

TABLE 1.2 Late nineteenth- and early-twentieth-century expeditions to Antarctic mainland

Years	*Expedition name*	*Leader*
1897–99	Belgian Antarctic Expedition	Adrien de Gerlache
1898–1900	British Antarctic Expedition (*Southern Cross* Expedition)	Carsten Borchgrevink
1901–04	British National Antarctic Expedition (*Discovery* Expedition)	Robert Falcon Scott
1901–03	First German Antarctic Expedition	Erich von Drygalski
1901–03	Swedish Antarctic Expedition	Otto Nordenskjöld
1902–04	Scottish National Antarctic Expedition	William Speirs Bruce
1903–05	Third French Antarctic Expedition	Jean-Baptiste Charcot
1907–09	British Antarctic Expedition (*Nimrod* Expedition)	Ernest Shackleton
1908–10	Fourth French Antarctic Expedition	Jean-Baptiste Charcot
1910–12	Japanese Antarctic Expedition	Nobu Shirase
1910–12	Amundsen's South Pole Expedition (Norway)	Roald Amundsen
1910–13	British Antarctic Expedition (*Terra Nova* Expedition)	Robert Falcon Scott
1911–13	Second German Antarctic Expedition	Wilhelm Filchner
1911–14	Australasian Antarctic Expedition	Douglas Mawson
1914–17	Imperial Trans-Antarctic Expedition	Ernest Shackleton

parties had to face in extreme environmental conditions, causing the death of some individuals and lacking mechanical developments in transport and communication that were subsequently available. It is worth considering that in the late nineteenth and early twentieth centuries the coasts of the Antarctic mainland became largely mapped and that different areas of the interior were explored, reaching the Geographical Pole in 1911. This period has been usually portrayed as foundational to Antarctic history to the detriment of previous human encounters with the region, such as sealing exploitation. Stories of the so-called 'heroic era' tend to be deeply infused with nationalism, and they have found appeal for their romantic views of the region and the figures of explorers. All of this has helped produce a particular dominant narrative of Antarctic history, having an impact on research interests and the nomination of HSM.

The expeditions of the late nineteenth and early twentieth centuries lived either in their ships or in prefabricated huts ashore (Pearson 1992; Foscari 2021b). Seven hut sites survive from that era: Scott's at Hut Point and Cape Evans, Nordenskjöld's at Snow Hill Island, Shackleton's at Cape Royds, Mawson's at Commonwealth Bay, and Campbell's (from Scott's 1910 to 1913 expedition) and Borchgrevink's at Cape Adare. While the experience of life at the expedition bases is in part outlined in the published works that resulted from them, the materiality of the experience, the understanding of the objects, their meanings and the human behaviour that led to their distribution and use, is hugely enhanced by the archaeological excavation of the remains. Their archaeological study is outlined in Chapter 2.

The period post–World War I saw increased exploration of Antarctica by air, with the expeditions of Lincoln Ellsworth, Hubert Wilkins, and Richard Byrd in the 1920s and 1930s, with some related land-based stations such as Little America being established. Land-based occupation accelerated during World War II, with both Britain and America establishing bases, and immediately after the war a number of ambitious government-sponsored expeditions from the United States, Britain, Norway, Sweden, France, Australia, Argentina, Chile and the Soviet Union paved the way for the International Geophysical Year in 1957–58, and the subsequent development of the Antarctic Treaty in 1959. This late twentieth-century era of cooperative research, extending now into the twenty-first century, has seen the human footprint in Antarctica multiply exponentially, and with it the potential for future archaeology, despite efforts such as the Environment Protocol to reduce waste and remove redundant materials from the continent.

1.2 The material/textual context

There are relatively few first-hand accounts of sealing compared with whaling with its higher 'romance' factor and the interest sparked by literary works such as Herman Melville's *Moby-Dick*, from 1851. The existence of novels on sealing, such as James Fenimore Cooper's *The Sea Lions*, from 1849, remained almost unknown to the general public. Because of this, you will find reference to the same sources recurring through this book – among them John Nunn and Nathaniel Taylor on Kerguelen; William Phelps on Marion Island; Thomas Smith on South Georgia; Nathaniel Ames, John Davis, Robert Fildes, James Weddell and others in the South Shetlands; Amasa Delano everywhere; James Robinson and Joseph Fuller on Heard Island; Charles Goodridge on Iles Crozet (Delano 1817; Fildes 1820–21, 1821a, 1821b; Weddell 1827; Ames 1830; Goodridge 1839; Smith 1844; Clarke 1850; Phelps 1871; Taylor 1929; Stackpole 1955; Busch 1980b; Nash 2009). Many of them are described in a recent work by Robert Headland (Headland 2018a). Logbooks and other documents (including crew lists, outfitting lists, etc.) represent an alternative source of information (Stackpole 1955; Salerno et al. 2018; Salerno and Cruz 2019), although as unpublished records they need to be searched for at different archives, and manuscripts frequently demand transcription for study. After the sealers, there is a body of substantial books arising from government-sponsored scientific expeditions of the nineteenth century. Compared with sealing and early exploration works, there is a very large literature of first-hand accounts from the 'heroic era' expeditions, and a vast collection of secondary histories, which is added to by what seems like dozens more every year, to the point where this 20-year period a century ago has come to dominate the public perception of Antarctic history. Since the 1940s, the literature has been largely divided between scientific reports and travel or work memoirs. On top of this, the Antarctic has been an important source for fictional writing (Leane 2012).

This body of literature provides textural ways of seeing the past. It relies on the interests, observational powers and veracity of the writer, and there is a great range within each of those aspects. Often the writer has a direct interest in conveying a particular impression of events – they may be the ship's captain or owner stressing success or the seaman/sealer stressing hardships; the leader of an expedition, naval or civilian, stressing good leadership and unanimity of the party, but downplaying disunity or problems; the traveller or popular scientist stressing adventure and misadventures but understating everyday behaviour (Mota 2018). Published versions may be selective or self-censoring in what is drawn from the on-the-spot original manuscript, and they tend not to say much about everyday life. Unpublished logbooks of the sealing era also present their own biases, as they tend to focus largely on the exploitation of resources (including the movements of the vessels, the distribution of the sealing gangs, the amount of skins or oil taken) (Salerno and Cruz 2019) and provide comparatively less data on sealers' lives on the hunting grounds.

The strength, and the challenge, of archaeology is that it deals with the material remains of these human occupations. In some cases, such as many of the sealing shelters, there is little other evidence available on which to reconstruct how men lived and worked in these locations. The archaeological evidence can tell us something about those things that left a material reminder – what was taken ashore and not removed, how a living shelter was built, how local resources were used to augment food and equipment brought ashore, and sometimes about how life was lived. Archaeological information on the distribution and characteristics of sealing sites could be integrated with historical data on landings and the establishment of camps with an aim to study general hunting strategies over time (Salerno and Cruz 2019), while the potential existence of historical descriptions for specific camps that today represent archaeologically detected sealing sites could be discussed. In other cases, such as the 'heroic era' huts, there is a more direct dialogue that can be reconstructed between the material evidence of the site and its contents, and the textual interpretation of events. That dialogue sometimes reveals what isn't said in the written record – where the physical evidence can reveal actions that are either denied or overlooked in what the observer wrote, the everyday things that reflect behaviour in a harsh environment or that illuminate how well-designed plans are modified in their execution (Figure 1.4). These ideas will be taken up in Chapters 3 and 4.

In an entirely modern historical context such as Antarctica, the combination of the material and the textual evidence and the dialogue between them are necessary to achieve an even partial understanding of the past. It is important to stress that researchers can have multiple perspectives on the ways that the relationship between historical and archaeological evidence might be established (Senatore and Zarankin 1996). Some researchers still consider that historical

FIGURE 1.4 Absolute Magnetic Hut ruin at Mawson's Huts, found to contain iron fixings, counter to the documentary record (Photo by Michael Pearson 1986).

archaeology may represent an auxiliary discipline to history. As a result, they understand historical information as the guide providing reference to interpret archaeological findings. Within this view, material remains are mainly used to confirm (and less likely to reject) documentary data. Meanwhile, some other researchers claim that historical archaeology has an independent status, defining its own agenda and problems of investigation. Within this framework, historical and archaeological evidence are studied independently, demanding particular approaches to deal with their nature. Eventually, information obtained from historical and archaeological evidence might be integrated, giving the opportunity to raise questions on the existence of consistencies, inconsistencies, presences and absences, among others (Galloway 2006).

1.3 Working in a unique geo-political environment

The unusual characteristics associated with archaeological work in the Antarctic and the sub-Antarctic start with the environment. Natural formation processes are key to understand the forms that the archaeological record takes in the region (Schiffer 1987). Sites located on the Antarctic and sub-Antarctic islands need to survive the activity of wildlife, the instability of slopes or shores, soil freezing and thawing, rain and snow, among others. The actions of elephant seals and penguins are worth noting. Wallowing is a means for elephant seals to remove moulting skin by rubbing and to make comfortable resting hollows. The presence of timber and other hard materials associated with places where humans lived and worked provides abrasives that attract wallowing, and as a result, many sites are disturbed or destroyed by elephant seals. While this is true for the South Shetland Islands, elephant seal colonies are much more abundant and greater in size on the sub-Antarctic islands. Penguins don't so much erode sites as bury them, many sites located on or close to penguin rookeries can have major accumulations of guano and moulted feathers obscuring the remains or totally burying them. Researchers do not only consider these processes for understanding the archaeological record, but they sometimes have to deal with animal activity while performing fieldwork.

Soil instability and animal action should also be noted. In the sub-Antarctic islands, following the very great loss of seals during the eighteenth and nineteenth centuries, coastal vegetation such as tussock (*Poa flabellate*) on the South Georgia archipelago took over areas formerly kept clear by animal movement. By the late twentieth century, many sealing sites were partially buried by regrowth tussock – the same occurred on Heard Island (Figure 1.5); this, to some degree, protected the archaeological remains from some forms of damage. With the massive recovery of fur seal numbers on South Georgia to an estimate of three times the pre-sealing population, however, the regrowth tussock has been stripped away in many locations, leading to severe erosion and animal damage to the sealing site remains, and disruption by other species (e.g., Bonner 1985; Cordeiro

FIGURE 1.5 Try-pot buried beneath tussock at Skua Beach, Heard Island. Many sub-Antarctic islands have very active depositional and erosional forces hiding or exposing sealing artefacts (Photo by Angela McGowan 1986).

2019). Another major disruptor of coastal vegetation has been introduced herbivores: reindeer and rats in the case of South Georgia; and rabbits, cats and rats in the case of Macquarie Island being the most serious. Major eradication programs have seen the removal of these animals in the last decade but not until after substantial damage had been done. A number of sealing sites in South Georgia are known to have been impacted by erosion, including the try-works site at Elsehul. A number of sealing and penguining sites on Macquarie Island have been partially buried by landslide, while others have been further exposed by erosion.

Sea-level fluctuations, due to either absolute sea-level changes or storm surge activity, have a significant impact on coastal sites, as well as freezing and thawing processes and glacial outflows that create seasonal lagoons and streams. On the South Shetland Islands, for instance, stratigraphical evidence has shown that some sites of the sealing era could have been covered by water at some point of their history. As these sites are very close to sea level, storm surge is the most likely cause. The inability to locate at least one previously recorded site located close to sea level suggests that sites are at risk of destruction by storm surge and potentially sea-level change. Soil freezing and thawing are especially worth mentioning on the South Shetland Islands, as these processes can affect the integrity of material remains and potentially translocate them (the probable impact of permafrost degradation is attributed to the 1902 Snow Hill Island hut in Chapter 2). Archaeologists working in the Antarctic environment also deal with the effects of high winds eroding and scattering material remains. Snow accumulation also represents particular challenges for researchers. In the case of the explorers' huts, for instance, some structures have become filled with snow, and excavation has demanded special techniques for recovering and conserving the artefacts, as described in Chapter 3.

While natural formation processes are relevant in the Antarctic and the sub-Antarctic islands, cultural processes should not be underestimated (Schiffer 1987). The action of scientists working in the region and the presence of tourists can sometimes disturb archaeological sites. On the South Shetland Islands, researchers other than archaeologists have sometimes established their camps on sealing sites or have removed rocks or whale vertebrae to use them somewhere else without even knowing the historical significance of these places (Pearson et al. 2010). Tourists have become a regular presence at some locations where sealing sites have been found and even more so in areas where no site surveys have been carried out. While tourists are encouraged not to destroy or remove materials of the environment, the lack of formal recognition of the sealing sites as part of the historical heritage of the region makes it difficult to implement effective protection measures (Senatore and Zarankin 2012). This situation is made worse by the humble materiality of the sites which often makes them invisible to the untrained eye.

Some other major factors influencing polar archaeology are the financial and logistical problems. It costs a lot of money to raise a conservation or archaeological expedition to Antarctica or the sub-Antarctic islands, and the proposal has to have the support of a managing nation to issue permits for the work and usually to provide the logistics of air and/or ship access. Researchers involved in these expeditions have to invest substantial periods of their time, often several months, to travel and field camp living with flexible timeframes for return, due to the 'uncertainty factor' that applies to all Antarctic logistics imposed by unpredictable weather and ice conditions, and the juggling of the needs of multiple scientific expeditions being deployed and supported during the same season. In most cases, commercial transportation is not feasible, and even if it is, it is very expensive and subject to the same 'uncertainty factor'.

Conservation work in Antarctica is limited to places that are recognised either in the list of HSM maintained by the ATS or within the boundaries of national scientific stations. In both cases, these sites are limited to those places that reflect stories that nations wish to tell about their role in the history of Antarctic exploration, science or Treaty activity. The list of HSM has just 90 places across the whole continent. It is the only formal mechanism for identifying sites of heritage value in Antarctica, and it is not based on a systematic identification and assessment of the heritage values of sites but is a mechanism for individual nations or nations in combination to identify sites important to them. While the Treaty outlaws acts that assert or support a claim to territorial sovereignty, the HSM list could be interpreted as a mechanism that in a small way does exactly that – it allows nations to nominate not just historic sites but recent monuments to national presence, effectively marking their legitimacy as current or future claimants to resource rights of 'scientific ascendency' on the continent. HSM are nominated by national parties, and no nominated HSM has ever been rejected by a Treaty meeting. Efforts have been made over time within the ATS to tighten up the HSM assessment process through the adoption of assessment criteria and the provision of more detailed information about each site, but given the consensus basis for decision making the process of reform is glacially slow with only small increments over the last 30 years.

Sites nominated to the HSM list include the obvious icons of Antarctic exploration such as the huts and other sites and monuments related to the 'heroic era' explorers of the late nineteenth and early twentieth centuries, the early scientific and exploratory sites of the later twentieth century and the Treaty era, graves associated with those, and monuments to national activities. All of these sites help produce and reproduce historical narratives where national authorship is clear. These narratives bring to the fore events and characters that are presented as milestones for a nationalised history in Antarctica and are seen as a proof of national commitment to the territory. Furthermore, along with national interests, these sites are considered to stand for the highest values proposed by the ATS, including the development of activities that contribute to the advancement of science and knowledge in a peaceful and collaborative environment. These narratives show respect to an ideal of nature conservation, making no reference to national interests surrounding the exploitation of resources, a subject that nonetheless keeps on arising – the Protocol on Environmental Protection to the Antarctic Treaty being signed in 1991 as a way to regulate human presence in the continent and seek at least a transitory halt on national claims for the exploitation of mineral resources.

Because the nomination of sites to the HSM list rests entirely with nations, sites that do not raise specific national interest and do not seem to represent the values of the ATS do not get nominated. Sealing sites cannot be easily associated with any single nation's activities. Therefore, they are not able to produce or reproduce stories of national achievement, but only some sort of multi-authored historical narrative. As such, sealing sites are hard to link to specific events and characters, or national involvement in Treaty-related objectives, providing instead evidence of historical processes and the traces left behind by multiple and anonymous people. Although sealers' presence was essential to the early exploration of Antarctica, and the number and distribution of sealing sites on the South Shetland Islands reflect the success of this activity, the materiality of campsites stresses their focus on exploitation. As a result, sealing sites do not sit comfortably with the values of peace and science that are icons of the ATS and have difficult or embarrassing associations in light of the ATS's emphasis on nature conservation. For all these reasons, sealing sites are both distasteful and not acknowledged as signifiers of responsible national involvement. The end result is that there are no sealing sites on the HSM list despite their obvious international historical significance.

However, it has still been possible for university or museum-funded archaeologists to get national logistical support for expeditions to research sealing sites, particularly the ongoing international research project 'Landscapes in White' based in Brazil, and earlier related programs supported by Chile and Argentina, described in Chapters 3 and 4. But none of these sealing sites have been the subject of formal conservation programs, and remain at risk.

A recent development in guidelines relevant to archaeological work is the adoption by the International Council on Monuments and Sites (ICOMOS) in 2022 of the Antarctic Archaeology Guidelines, developed by the International Polar Heritage Committee (IPHC), one of ICOMOS's International Scientific Committees (IPHC 2022). These guidelines promote the open-science basis of Antarctic research, draw on existing international standards in detailing best practice and tailor its guidance in the light of the environmental and logistical circumstances of Antarctica.

In the sub-Antarctic islands the governance is in the hands of specific nations, and the islands are managed more or less as national parks might be managed, with surveys of all cultural sites regardless of their historical origins, and conservation works carried out as and when resources and mounting risk factors dictate. Some of this work is outlined in Chapter 2.

Within Antarctica, the activities of the citizens of a country and the Antarctic program of each nation (as a Treaty Party) are regulated by that nations' laws and requirements, within the overarching set of policies and protocols agreed through the ATS for Antarctic environmental protection and management. As a result, there are varying processes and standards applied in relation to the study and management of cultural heritage, just as there are between countries. Some countries, for example, have adopted ICOMOS principles within their national legislation and administration processes, with active heritage survey and listing programs, and while some apply these approaches in Antarctica to varying degrees, others have not in practice extended their domestic heritage controls to the Antarctic context. Within the realm of the ATS requirements, protective measures for cultural heritage are only envisaged within those places listed as HSM, Antarctic Specially Protected Area (ASPA) and Antarctic Specially Managed Area (ASMA). As discussed earlier, this sets up a nationally-defined and limited set of heritage places, and even if domestic legislation requires such things as heritage survey before development and value-based heritage listing programs, these will not necessarily be implemented within Antarctica.

In the absence of uniform approaches, and in light of the resulting substantial gaps in protection and management control, a number of organisations have taken steps to define and promote standards relevant to their community of interest or in an attempt to improve the overall understanding of the needs for best practice. Examples relevant to the cultural heritage are IAATO (the International Association of Antarctic Tour Operators), which has the primary objective to advocate and promote the practice of safe and environmentally responsible private-sector travel to the Antarctic, with operations that have no more than a minor or transitory impact on the Antarctic environment; and the IPHC (International Polar Heritage Committee – an International Scientific Committee of ICOMOS), which brings together cultural heritage conservation expertise from around the world to promote and advise on the standards of conservation in polar regions. The IPHC has developed standards-related guidelines and advice papers, such as the Antarctic Archaeology Guidelines adopted by ICOMOS and the Antarctic Underwater Material Cultural Guidelines developed in collaboration with the International Committee on Underwater Cultural Heritage of ICOMOS.

The following chapters elaborate on the scope and nature of archaeological work in the Antarctic and the sub-Antarctic, undertaken within the unique governance environment described here.

2

THE HISTORY OF ARCHAEOLOGICAL INVESTIGATIONS IN ANTARCTICA AND THE SUB-ANTARCTIC ISLANDS

A number of publications have looked at the history of archaeological work in Antarctica (Harrowfield 2005; Pearson 2011; Pearson and Senatore 2014; Zarankin and Salerno 2017; Senatore 2019). Much of the archaeological work from the 1970s onwards has been related to the understanding and conservation of historic sites and buildings. These include the most obvious – those given a privileged place in terms of historical 'importance' by the historical literature: the 'heroic era' exploration huts. Archaeological study of historic sites, however, has also included nineteenth and twentieth-century scientific stations, shipwrecks, whaling stations and sealing sites (Table 2.1).

Two major parallel avenues of archaeological work can be seen in the literature. The first is the use of archaeology at exploration hut sites as a tool to gather information about the material aspects of their occupation, to better understand the structure and construction history of the huts, and as a controlled approach to removing ice and uncovering building fabric and artefacts as a part of heritage conservation programs. This is usually undertaken as part of government or NGO-supported efforts to conserve historic sites identified in the literature and usually designated within the Antarctic Treaty System (ATS) as Historic Sites and Monuments (HSM). A closely related motivation is the survey of sites for management purposes (especially in the sub-Antarctic islands) or as part of environmental impact assessments before development or environmental clean-up works.

The second use of archaeology is as a research tool, through survey and excavation, to study, record and understand human activities related to research bases, whaling stations and sealing sites in both the sub-Antarctic islands and the South Shetland Islands, largely or wholly independent of conservation objectives. This has been undertaken either as part of academic research programs, as in the South

DOI: 10.4324/9780429201257-3

Shetland Islands, or as a part of broader programs by Antarctic program managers to define the natural and historic resources that need to be protected and managed by them, as in the sub-Antarctic islands, already referred to. Some of the whaling sites and most of the sealing sites involved in these studies have not been previously identified or detailed in the popular historical literature and have only been located and described through archaeological survey.

TABLE 2.1 Timeframe for archaeology in Antarctica and the sub-Antarctic islands

Date	*Location/project*	*Source*
1957–58	FIDS observation of sealing sites on Livingston Island during geological and biological surveys. No professional archaeological input. All artefacts recovered disposed of by order of the then Director of SPRI.	Simpson (1959); O'Gorman (1963); Hobbs (1968); Lewis-Smith and Simpson (1987)
1960–61	Scott's Cape Evans and Shackleton's Cape Royds huts restored. Ice dug out and many artefacts disposed of. No professional conservation or archaeological input.	Quartermain (1961, 1963); Ritchie (1990); Harrowfield (2005)
1961	Borchgrevinck's Cape Adare hut ice dug out to provide emergency shelter to a party.	Harrowfield (2005)
1963–64	Scott's *Discovery* Hut at Hut Point restored. No professional conservation or archaeological input. Ice dug out and many artefacts disposed of. Middens bulldozed. Such work continued through to 1971.	Ritchie (1990); Harrowfield (2005)
1974	Borchgrevinck's Cape Adare hut gets 'preventative maintenance'. No professional conservation or archaeological input.	Ritchie (1990); Harrowfield (2005)
1977	Scott's Cape Evans Hut: David Harrowfield excavated a test trench. Harrowfield had some archaeological field training.	Harrowfield (1978)
1978	Mawson's Huts: ANARE work on hut by Rod Ledingham party, no professional archaeology, but artefacts location documented and left in hut.	Ledingham (1978); Harrowfield (2005)
1979–80; 1980–81	Excavation of artefacts from frozen ground at Nordenskjold's Snow Hill Island Hut, as ice dug out. Santiago Comerci (historian, 1922–2020). Ricardo Capdevila (lawyer/historian, 1936–2008) worked at site 1981–2007	Comerci (1981, 1983); Capdevila and Comerci (1986)
1981	Ronald Lewis-Smith examination of sealing sites on Livingston Island during geological and biological survey. No professional archaeological input but a competent ground survey.	Lewis-Smith and Simpson (1987)

1982	Documentation of Borchgrevink's Hut and study of salt deposition and wind impacts.	Harrowfield (1982, 1984, 2005)
1982	Preliminary survey of sites on Macquarie Island (archaeologist Martin Davies).	Davies (1982)
1983–95	Archaeological survey and excavation of sealer campsites at Admiralty Bay, Cape Shirref and Desolation Island in the South Shetland Islands by Rubén Stehberg, Chile. First professional archaeological work in Antarctica.	Stehberg (1983a, 1983b, 1997); Stehberg and Cabeza (1984, 1987); Stehberg and Lucero (1985a, 1985b, 1995, 1996)
1984	Strategic Plan for Ross Sea sites adopted, with requirement for archaeological involvement.	Turner and Harrowfield (1984)
1984–85	Mawson's Huts: Project Blizzard excavations, artefact scatter recording (Estelle Lazer). First professional archaeological work on huts in Antarctica.	Lazer (1985); McGowan, A. (1987, 1988)
1985–86	Mawson's Huts: Project Blizzard excavations (Angela McGowan). First-time archaeologist and conservator worked together in Antarctica (Janet Hughes).	McGowan (1987, 1988); Hughes (1986)
1986	Australian Antarctic Historic Sites and Monuments Advisory Committee formed by AAD.	Harrowfield (2005)
1986–87	Mawson's Huts: hut roof deflection measurements and ground artefact movement study (Michael Pearson for AHSMAC).	Pearson (1986)
1986–87	Archaeological survey of Heard Island (Estelle Lazer and Angela McGowan).	Lazer and McGowan (1990); Hughes and Lazer (2000); McGowan (2000); McGowan and Lazer (2018)
1986–88	Survey and excavations of sealing and penguining sites, Macquarie Island (Karen Townrow).	Townrow (1989, 1990)
1986–2007	First NZ professional excavations at Cape Evans Hut. (Neville Ritchie, Simmons and Cross). Followed by seasons of documentation and excavations at Cape Evans (1988–89 to 2008), Butter Point (1988), Cape Adare (1990, 2003), Cape Crozier (1992, 2002, 2003), Granite Harbour (2002, 2003) and Cape Royds (1994–2006). No major archaeological excavation work since 2008.	Ritchie (1988, 1889a, 1989b, 1990, 2005a, 2006); Harrowfield (1991, 2005); Fyfe (1990, 1992); Ritchie and Fyfe (1995); Fastier et al. (2006)
1987	Antarctic Heritage Trust, NZ, created, and new management plan 1990 for Ross Sea sites required archaeological input.	Cochran (1990); Harrowfield (2005)

(Continued)

TABLE 2.1 (Continued)

Date	*Location/project*	*Source*
1988–89	Documentation of Wilkes Station (Linda Clarke and Elspeth Wishart).	Clark and Wishart (1989)
1989–90	Documentation of Old Casey Station (Clark and Wishart).	Clark and Wishart (1990)
1990–97	Industrial archaeological survey and mapping of whaling stations at Husvik, Stromness, Grytviken, Leith Harbour, Prine Olaf Harbour and Ocean Harbour on South Georgia (by Bjørn Basberg, Dag Nævestad, Gustav Rossnes and Gisle Løkken). Followed up with 3D digital recording by Russell Gibb and Daniel McCurdy from 2012.	Nævestad et al. (1996); Basberg (2004); Gibb and McCurdy (2013)
1990–91	Archaeological survey of a whaling station, and Base B of Operation Tabarin, at Deception Island (Louwrens Hacquebord).	Hacquebord (1992)
1991–92	Survey and selected artefact removal at US East Base, Stonington Island. (Robert Spude and Catherine Holder Spude) followed up by Noel Broadbent.	Broadbent et al. (1992); Spude and Spude (1993); Parfit and Kendrick (1993); Broadbent and Rose (2002)
1991–92	Survey of human impacts at whaling and scientific station site at Whaler's Bay, Deception Island, South Shetland Islands (Louwrens Hacquebord).	Hacquebord (1992)
1993	Survey and assessment Mawson Station (Linda Clarke and Elspeth Wishart).	Clark and Wishart (1993)
1993–94	Survey of sites on Kerguelen and Iles Crozet to create an inventory by 2000 (Jean-François Le Mouël).	Le Mouël (1994, 2004)
1993–94	Land and underwater survey to try to locate remains of the Spanish frigate San Telmo, lost in 1819. The project, led by Manuel Martín-Bueno, failed to find definitive evidence.	Martín-Bueno (1995a, 1995b, 1996); Martín-Cancela (2018)
1994	Noel Broadbent surveyed the airfield and 1957 campsite at Marble Point, Ross Sea.	Broadbent (1994)
1994–96	Base A of Operation Tabarin (1944), Port Lockroy, worked on, and partly excavated, by non-archaeological personnel for BAS.	Burkitt (1996); Cochran (1990)
1995–96	Excavation of 1820s sealer shelter 'Hope Cottage', Kerguelen Island (Le Mouël).	Le Mouël (1996, 2004)
1995–96	Mawson Station heritage assessment (Sam Rando and Martin Davies).	Rando (1995); Rando and Davies (1996b)

1996-date	Surveys and excavations on Byers Peninsula, Livingston Island, South Shetland Islands, by Andrés Zarankin and Ximena Senatore, and ongoing multi-national 'Landscapes in White' research project led by Zarankin.	Zarankin and Senatore (1999a, 1999b, 2005, 2007)
1996	Documentation of Davis Station (Sam Rando and Martin Davies).	Rando and Davies (1996)
1996–98	Survey of whaling artefacts around former whaling anchorage at Admiralty Bay (Piotr Kittel et al.).	Kittell (2001)
1997–98, to 2015–16	Mawson's Huts Foundation created in Australia in 1996, work at Mawson's Huts began in 1997 (archaeological component Estelle Lazer). Subsequent expeditions included archaeologists Anne McConnell, Jody Steel, Stirling Smith, and a range of conservators, notably Ian Godfrey. No archaeological program from 2011 to date (2020).	Lazer (1998); et al. Mawson's Huts Foundation newsletters.
2000–01	Excavation and restoration of whaling station buildings, Port Jeanne d'Arc, Kerguelen Island, (Jean-François Le Mouël).	Le Mouël (2004)
2000–05	Excavation of William Spears Bruce's 1903 weather station in the South Orkney Islands, Omond House (Veronica Aldazebal).	Aldazabal (2007)
2001–08	Surveys and excavations on Fildes Peninsula, Rugged Island and Byers Peninsula, in the South Shetland Islands (Rubén Stehberg, Michael Pearson).	Stehberg (2002, 2003, 2004); Pearson et al. (2006, 2008, 2009)
2003	Archaeological survey of Auckland Island for sealing and other sites (Nigel Prickett).	Prickett (2009)
2006	Excavation and removal for relocation above wave erosion of try-works at Vallée des Phoquiers, Iles Crozet (Paul Courbon and Jean-François Le Mouël).	Le Mouël (2006); Le Mouël et al. (2013); Courbon (2012)
2006–07	Excavation of Observation Bay scientific station, Kerguelen Island (1873–1903, British and German) (Alexandra Barbot).	Courbon and Le Mouël (2009); Lüdecke (2004)
2007–2014	Excavation related to investigation and conservation of the Scott and Shackleton huts in the Ross Sea, by the Antarctic Heritage Trust (NZ).	Skinner (2014)
2011–13	Survey and excavations of survivor campsite and shipwreck site of the sealer Solglimt on Marion Island, Prince Edward Islands, by Tara Van Niekerk and others.	Graham (1989); Van Neikerk (2016); Boshoff et al. (2015); Cooper et al. (2016)
2017	AHT NZ completes the main program of huts conservation at the Scott and Shackleton huts.	www.nzaht.org

(Continued)

TABLE 2.1 (Continued)

Date	*Location/project*	*Source*
2017	1957 Hillary Hut restored by NZ AHT. No archaeological component (the hut having been relocated in 1989).	Watson and Ussher (2017)
2017	Conservation work continues on Nordenskjold's Snow Hill Island Hut (ongoing since 1979), first archaeologist Victoria Nuvialla in 2017.	Victoria Nuvialla pers. comm.
2017	NZ and UK AHT joint project conserving British bases at Stonington Island, Port Lockroy and Horseshoe Island undertaken with no archaeological component.	Websites for the UKAHT and NZAHT
2018	Archaeology strategy being planned for Cape Adare huts by AHT (NZ).	Lizzie Meek pers. comm.
2019	South Georgia Archaeological Project commenced by the South Georgia Heritage Trust, to survey and excavate sealing and whaling sites on the island. Marcus Brittain lead archaeologist for Cambridge Archaeology Unit.	Neil (2019); Cordeiro (2019)
2020	Archaeological program commenced at Nordenskjøld's Snow Hill Island Hut, and the Hope Bay and Paulet Island shelters, led by Dag Avengo and Pablo Fontana.	https://meltinghistory.org

2.1 The archaeology of exploration and science – the 'heroic era' huts

The exploration of the sub-Antarctic islands and the process of defining the extent of an Antarctic continent started in the eighteenth century, while human landings in Antarctica itself, and its scientific study, started in the early nineteenth century (see Chapter 3). This early interaction with the continent was within the context of the ongoing scientific enlightenment in the West, and while the primary incentive for the first human activities on the land was the exploitation of sea mammals, many of the participating shipping owners and captains were deeply involved in personal scientific endeavours and were members of the early learned societies. The gathering of natural history information was both personally satisfying and helped build social status in the urban home society. Therefore, a number of sealers in the 1820s made written observations or collected natural history specimens.

Dr. James Eights, often credited as Antarctica's first scientist, described the South Shetland Islands and made collections there as a member of a private American sealing and exploration expedition in 1829–31. The British Naval Expedition to the South Atlantic commanded by Henry Foster between 1828 and 1831 included the survey of the South Shetlands archipelago, taking special note of Deception Island. The French, British and US naval explorations of the 1830s and 1840s (d'Urville, Wilkes and Ross) were underpinned by scientific objectives, in particular relating to magnetic research which both advanced the understanding of the nature of the planet and promised to greatly assist global navigation (Fogg 1992; Mawer 2006; Pearson et al. 2020). These early scientific endeavours were predominantly ship-based, and none left material evidence at sites in Antarctica that has as yet been identified. It was not until the end of the nineteenth century, the start of the period dubbed the 'heroic era', that scientists established shore bases where huts were built and artefacts were abandoned, intentionally and unintentionally, creating the material evidence for later archaeological study.

Between 1897 and 1917, some 15 major expeditions operated in Antarctica. Only one of them (the Japanese expedition under Nobu Shirase, 1910–12) was not European, while the United States did not participate at all, their 'heroes' heading for the North Pole instead. This explosion of expeditions to the Antarctic mainland was aimed at carrying out scientific and geographic research for the 'benefit of mankind' (e.g., Larson 2011; see Pearson et al. 2020). However, this was also a period of intense national rivalries and imperial aspirations, and as Adrian Howkins puts it, the Antarctic exploration expeditions were 'saturated by the politics of imperialism' (Howkins 2017: 33). This was not an imperialism based in territorial colonialism, in the traditional sense of taking control of a region by establishing colonies and settlers. The idea of physically colonising Antarctica was not seriously contemplated, except in the case of sealing entrepreneur James Byers unsuccessful attempt to get the United States to annex the South Shetland Islands in 1820 (Bertrand 1971: 34). Nor was it a case of trade imperialism of the kind that extended a massive Portuguese trading empire in the seventeenth and eighteenth centuries, and those of the Netherlands, France, Spain and Britain right into the twentieth century.

In the case of Antarctica, imperial influence was pursued through the demonstration of scientific and geographic 'mastery' of the Antarctic environment – not physical ownership but moral and intellectual control. Those countries that could generate major geographical and scientific knowledge and authority, and could ensure that the names they gave to places they 'discovered' were adopted by others, could extend their imperial reach by demonstrating an intellectual and moral dominance over parts of Antarctica and the surrounding oceans. This extended into the twentieth century through legal influence over whaling, and then in the post–World War II period when membership of the ATS was based on demonstrated scientific activity. Adrian Howkins (2017: 8) convincingly portrays this as imperialism by the demonstration of 'environmental authority':

> the assertion that the production of useful scientific knowledge about an environment helps to legitimate political control over that environment. The concept of environmental authority therefore combines the idea of doing better science, which does not necessarily involve political control

> with some sort of political claim to a territory, which does not necessarily involve scientific research.

It is notable today that this trend continues, with just 29 of the world's 195 countries exercising sole control over the continent through the Antarctic Treaty, justified by their active science programs (e.g., Dodds 2021).

Contemporary and subsequent histories of this period commonly see this 'heroic era' as the first meaningful human contact with Antarctica, ignoring or dismissing the sealing discoveries and occupation of the 1820s and later sealing, whaling and exploratory activities throughout the nineteenth century. It is images of the 'heroic era', and the names of explorers like Scott, Amundsen and Shackleton, that first enter the popular mind when 'Antarctica' is evoked (though this is perhaps being replaced with thoughts of the impacts of climate change in more recent times). The expeditions, and particularly those from Britain, were envisaged and written about as high adventures pitting men (and only men) against the elements to conquer nature in the sublimely hostile Antarctic environment: the glory of empire without the political and moral odium that might come with the subjugation of native peoples that was a customary practice at the inhabited regions of the world (see Bloom 1993; Howkins 2017). The perception of this period as a particularly masculine undertaking of *Boy's Own* derring-do is reflected in the name given to it in the 1950s by British writers – 'the heroic era' (Hince 2000: 167–168). While exploring parties of the late nineteenth and early twentieth centuries were made up of men, historical information and research studies show that some women requested to take part in those expeditions, though being rejected. Furthermore, as described in the Introduction, women have their own undeniable place in Antarctic history from early moments, defying male-centred preconceptions.

Beau Riffenburgh (quoting John McKenzie 1992) has outlined the key elements of the heroic myth: an exotic location; the supposed personal qualities and heroic characteristics of the individual, full of energy, stamina and decisive action; the martyrdom of the hero (noting that the Americans preferred their heroes alive, while the British preferred them dead); and the creation of heroic icons of the myth, interpreted as justification of the expansion of empire and economic reach, embodying the collective will and hopes of the governing elite, and offering guidelines for personal and national ascendancy to a new generation (Riffenburgh 1994: 6). All of these elements are present in the stories and images of the 'heroic era' and to a degree in subsequent eras in Antarctic history, and indeed still infuse the rhetoric of the Antarctic Treaty Consultative Party meetings: the guidelines for the listing of HSM, for example, adopted in 1995 and reiterated in guidelines through to 2018, include as one of the criteria for listing a place that it has 'a particular association with a notable feat of endurance or achievement'. In other contexts, Ernest Shackleton, in particular, is still evoked regularly as the *beau ideal* of the decisive leader in conditions of extreme difficulty, to imbue youth and the would-be leaders of industry with the spirit of the 'heroic' leader. No sealer gets that accolade, no matter how intrepid or successful as a leader of men or an expedition.

The sealing era has left few examples of publicly recognised 'alpha males'. A very small number of sealing captains have had passing or continuing fame, such as Nathaniel Palmer and James Weddell, but these are in general linked to claims of 'discovery' important in later national narratives. The ordinary sealer, and indeed sealing master, is either un-named or if named little documented beyond that recognition. It has been argued that this is in part because of the particular position of sealing in the economic and social hierarchies of the early nineteenth century (Zarankin et al. 2018). For one thing, whaling, the close relative of sealing, was in both Britain and the Unites States a more economically important industry than sealing. Whaling also was seen as being more 'noble', pitting brave men against the often 'malevolent' leviathan, and requiring learnt skills of a much higher order than those needed simply to club a 'defenceless' seal. Whaling therefore had a higher social status, and this is reflected in the telling of community and national stories. Whaling is also a more palatable topic for modern museums, as its presentation can be counterpoised with the story of the struggle to 'save the whales', a narrative that is not so easily told about sealing. The expedition leaders of the 'heroic era', on the other hand, had the benefits of representing their nation (or a leading industrialist), killing only for food (for them or their dogs) or as scientific specimens, receiving press accolades, and writing their own best-selling narratives.

The archaeological evidence associated with the huts occupied by the explorers is much more prosaic than the portrayal of the expeditions in historical and contemporary literature. While you can demonstrate physical hardship and difficult living conditions through material remains, you can't read 'heroic' into those remains – 'heroic' is a storied cultural response, not a self-evident characteristic of artefacts. As an example, the living conditions faced by the sealers of the early nineteenth century were often much harsher than those experienced at the explorer huts or field camps of the 'heroic era', but the label 'heroic' has never been applied to the sealing gangs, or used to characterise the often life-or-death interaction of sealing parties with the Antarctic environment. Considering this, many sealing logbooks contain interesting though not clearly considered references on the occurrence of accidents and the loss of men. One explanation may be that sealing was seen as a working-class industrial activity, carried out for making a profit, while exploration was seen as a middle-class expression of exemplary (and usually amateur) manly qualities, conducted for a greater purpose (science, glory, nationalism, etc.). A case of 'workers' vs. 'heroes'?

The archaeology of the 'heroic era' has largely focused on the conservation of the seven surviving expedition huts – Borchgrevink's at Cape Adare (1898–1900), Scott's at Hut Point (1901–04), Nordenskjöld's at Snow Hill Island (1901–03), Shackleton's at Cape Royds (1907–09), Scott's at Cape Evans (1911–13), Campbell's at Cape Adare (Scott's Northern Party 1911–12), and Mawson's at Commonwealth Bay (1911–14) – and the pioneer weather station in the South Orkney islands established by the Bruce expedition (1902–04). The notable exclusion of Amundsen's Bay of Whales hut from this list is because it was built on an ice shelf and has subsequently calved into the sea (Figure 2.1). Description of the archaeological work conducted at the explorers' huts will follow a geographical order, as these sites might be grouped into four regions: the Ross Sea area, Commonwealth Bay in King George V Land, Snow Hill Island in the Weddell Sea, and Laurie Island on the South Orkneys (see Pearson 1992 for description and Foscari 2021a: 614–639 for a series of plans).

2.1.1 Ross Sea region

The earliest conservation work in Antarctica was on the huts in the Ross Sea region, these being expedition bases built by the *Southern Cross* Expedition at Cape Adare (1898–1900), led by Carsten Borchgrevink; the *Discovery* Expedition at Hut Point (1901–04) and the *Terra Nova* Expedition at Cape Evans (1910–13), both led by Robert Falcon Scott; and the *Nimrod* Expedition at Cape Royds (1907–09), led by Ernest Shackleton. A hut built by the 'northern party' of the

FIGURE 2.1 Amundsen's 'Framheim' hut at the Bay of Whales, 1912. The hut was built on an ice shelf, and subsequently calved into the sea, so it is not available for archaeological study (Source: National Library of Norway, Creative Commons).

Terra Nova Expedition, led by Victor Campbell, on Cape Adare, has almost completely collapsed, and a decision was made not to attempt its conservation.

The *Southern Cross* Expedition worked on the Ross Sea area, as its accessibility has already been shown by the James Clark Ross expedition in 1841, and Cape Adare had been visited by the leader, Carsten Borchgrevink, on Henryk Bull's expedition in 1895. The party carried out scientific observations and established a base camp on Cape Adare, on the north-easternmost extremity of Victoria Land, being the first to over-winter on the Antarctic mainland in 1899. The camp included the erection of two structures, one used as a living hut and the other one as a store. These were the first architectural structures in Antarctica, as opposed to the rough shelters built by the sealers in the 1820s. After their wintering, the returned party sailed to the 'Great Ice Barrier' (the Ross Ice Shelf which represented a major obstacle at the time) and made the first journey there using dogs and sledges, travelling a few miles from the ship and establishing a Farthest South record in 1900. Scott's *Discovery* Expedition also worked in the Ross Sea area, being based at Hut Point, Ross Island, in McMurdo Sound, where they erected a hut in 1902. The expedition conducted scientific observations and discovered the Polar Plateau where the Geographic South Pole is located. Even though the expedition could not reach this location, it established a new Farthest South record.

Shackleton's *Nimrod* Expedition also sought to reach the South Pole and achieved multiple scientific goals. It was based on Ross Island as had been the *Discovery* Expedition, but the party landed on Cape Royds, where the men erected a hut in 1908. While one of the groups making up the expedition established a new Farthest South record when they tried to reach the Geographic South Pole (making the decision to return home in view of the extreme conditions experienced); the other one reached the location of the South Magnetic Pole. Scott's *Terra Nova* Expedition also pursued the Geographic South Pole, while it also carried out scientific research. The expedition was also based in a hut built in 1911 at Cape Evans on Ross Island, about 20 km from Scott's former base at Hut Point. While Scott's group reached the South Pole, it was a month after Amundsen's expedition made it, and the men died in their way back home. Another group of the expedition (the 'northern party' led by Victor Campbell), established a hut on Cape Adare where they spent the winter of 1911. Amundsen's *Fram* expedition was based on the Great Ice Barrier at the Bay of Whales, the southernmost point of open sea in the world, from where he drove dog sleds across the ice barrier and Transantarctic Mountains to the South Pole. Amundsen's hut *Framheim* subsequently calved into the Ross Sea, so no archaeological remains of the base survive.

The first 'conservation/preservation' work in Antarctica was carried out by sailors of HMNZ *Endeavour* and personnel of the New Zealand Scott Base in 1957–58. It consisted of 'repairing', removing some snow, and making a 'clean-up' of rubbish at the Scott and Shackleton huts (Harrowfield 2005: 8). Following the agreement of the Antarctic Treaty in 1959, New Zealand, in order to demonstrate its responsible management of the exploration huts under its control, decided to undertake work to restore them.

> It was agreed that the aim should be to restore the huts to their original appearance and to preserve them from further damage; only essential construction work was envisaged, and it was not desired to give the huts the appearance of museums.

In the 1960–61 season, a party led by historian Leslie Quartermain removed the accumulated snow that two-thirds-filled Scott's Cape Evans Hut with pick and shovel: 'The useless rubbish was drawn by sledge to the fast ice, where it either went down a tide crack or disappeared when the ice broke out some weeks later' (Figure 2.2). Shackleton's Cape Royds Hut was free of ice, but 'untidy inside and out': 'Obviously useless rubbish was burnt, while spare foodstuffs and other objects were carefully stock-piled and listed.' ... 'A careful selection and distribution of the material found inside and outside made it possible to re-create something of the original atmosphere.' ... 'Finally, the surroundings were thoroughly cleaned up...' (all quotes Quartermain 1961: 606). No archaeological expertise was involved, no sketch plans or measurements of the locations of the hundreds of artefacts excavated were made, and few photographs were taken (Harrowfield 2005: 8), though this was not unusual for what were still seen as 'recent' buildings at the time (only 50 years old). The practice of placing objects within the

FIGURE 2.2 Scott's Cape Evans Hut, 1910–12, with Mount Erebus beyond. The section in the foreground was the stables for ponies taken to haul sledges (Source: Flickr, Creative Commons).

huts, based on photographic or journal evidence, or on personal taste to recreate an imagined frozen moment in time (Figure 2.3), has been aptly named the 'archaeology of nostalgia' by María Ximena Senatore (2019).

Borchgrevink's Hut at Cape Adare (Figure 2.4) was also excavated to remove ice in 1961 to provide emergency shelter for a field party, and Campbell's Hut (a substantially collapsed building) had artefacts removed from it. This time at least there was a sketch plan made of the interior, but there was no professional archaeology and no photographs were archived (Harrowfield 2005: 9).

In 1964, the ice in Scott's *Discovery* Hut (Figure 2.5), located at Hut Point on the edge of the huge US McMurdo Station, was removed again with pick and shovel. One floor plan was made by an architect, but two adjacent small huts and a Ross Sea Party (1914–17) midden of equipment and domestic refuse was cleared away by a US bulldozer operator: '...in those early days, systematic archaeology was not even thought about, let alone considered necessary' (Harrowfield 2005: 9–10). As late as 1971, a large midden associated with the Ross Sea Party was cleared away outside the main entrance of the Cape Evans Hut, including sledging, scientific and photographic equipment, clothing and cans, by a New Zealand works party charged with 'cleaning up' the area (Harrowfield 2005: 10; Ritchie 2006: 40)). The result of the 'clean-ups' at the Scott and Shackleton huts was that few undisturbed areas with archaeological deposits or artefacts remained intact around the huts, and those that did were of limited extent (Ritchie and

FIGURE 2.3 Stables recreated outside Shackleton's Cape Royds Hut, an example of the 'archaeology of nostalgia'. Pony stables existed in this location, but had been dispersed and eroded. The box walls were recreated – what is original and what is not? (Photo by Michael Pearson 1992).

FIGURE 2.4 Borchgrevink's Hut at Cape Adare, 1899–1900. The stores hut is roofless. The first architectural structure to be built in Antarctica, as opposed to the rough shelters built by the sealers 80 years before (Source: Oceanwide-Expeditions, Creative Commons).

FIGURE 2.5 Scott's *Discovery* Hut, 1901–04, with McMurdo Station just beyond it. While of moderate use to Scott, this hut was critical to the survival of Shackleton's Ross Sea party in 1914 (Source: Tas 50, Creative Commons).

Simmons 1987: 21). The prospect for future research archaeology was thereby substantially reduced.

David Harrowfield, a manager at the Canterbury Museum (Christchurch, New Zealand) who had received some local training in archaeology in the 1950s, in 1977 started experimenting with methods for the removal of artefacts from ice, using black plastic to thaw frozen deposits and recording the depth and location of artefacts retrieved (Harrowfield 2005: 11). The following year the first strategy for the Ross Sea historic sites was written by surveyor Gerry Turner as a dissertation, which called for the involvement of professional archaeologists in the excavation of historic deposits. It was the basis for a strategic plan written by Turner and Harrowfield, adopted in 1984 (Turner 1979; Turner and Harrowfield 1984). In 1982, Harrowfield was again involved in the site planning and artefact inventory at the Borchgrevink and Campbell huts at Cape Adare, and made a preliminary study of salt deposition and wind effects on the remains (Harrowfield 1984, 1996, 2005: 12).

The management and conservation of the Ross Sea huts was formalised in 1987 with the creation of the Antarctic Heritage Trust (AHT), a New Zealand not-for-profit organisation with close links to the New Zealand and British government Antarctic programs, that took on the practical responsibility for the conservation of the historic sites in the Ross Sea region on behalf of the international community. The Historic Sites Management Committee (the precursor to the AHT) had already engaged professional archaeologists on the Scott and Shackleton huts for the first time in 1986–87, when Neville Ritchie and Alexy Simmons commenced what became a long-term archaeological program implemented by the AHT. A new management plan was written for the Ross Sea region (Cochran 1990), which identified 34 historic sites of conservation interest (see also Harrowfield 1988), and included a policy requiring the involvement of appropriate qualified archaeologists in their study.

The rationale for both conservation and archaeological investigations was seen to be that

> The huts, and the associated artefacts, monuments, and field depots are the tangible remains of the early expeditions. As such they play an important role in transmitting history, traditions, and values to succeeding generations. In addition, they are 'time capsules' of technological achievement, vital sources of knowledge about undocumented facets of the expeditions, and constitute an invaluable comparative data base for research now and in the future.
>
> *(Ritchie and Simmons 1987: 1)*

The archaeological work in the initial AHT seasons at the *Discovery*, Cape Evans and Cape Royds huts focused on the interpretation of structural and occupation history aimed at better informing the proposed structural repairs, the recording of fabric and artefact accumulations before removal/conservation, and the disposal of inflammable liquids and leaking tins. A series of test excavations outside Cape Evans Hut were related to plans to move provisions boxes and other loose artefacts (originally built as Bowers' stores annex) away from the walls of the hut, where they were causing impoundment of melt-water, damaging both the hut wall and the stores themselves. The stores boxes and individual artefacts (tins, bottles and other containers) were frozen into a compacted snow bank and bonded into the permafrost impregnated ground at its base. It was estimated that 50% of the plywood stores boxes forming the walls of the annex would disintegrate on removal, with the contents (c. 2,000 cans) being left exposed and in need of assessment and conservation action (Ritchie and Simmons 1987). Subsequent seasons saw further work at Cape Evans Hut on Bowers' annex by Ritchie and Nelson Cross, the removal of ice and artefacts from the stables annex by Ritchie and Roger Fyfe (Ritchie 1988, 1989, 2006; Fyfe 1990, 1992), and the removal of frozen piles of food cans and residues, and the last remaining ice from inside the hut (Ritchie and Fyfe 1995). Since the mid-1990s, most of the archaeological work at Cape Evans has been limited to removal of seasonal ice and snow to prevent or remove ponding of melt-water around the hut (Ritchie 2005).

Archaeological work in 1987–88 also included the survey and removal of a cache of artefacts at the 1910–13 Butter Point field camp near the Bowers Piedmont glacier, across McMurdo Sound opposite Cape Evans Hut. The Butter Point cache artefacts, and those collected from the Inexpressible Island field cache, which comprised material evidence of the field resources available to scientific field parties of the time, remained in good condition in the AHT container at Scott Base when seen by Ritchie in 2008 (Ritchie 1990a: 26; Ritchie 2008). Some excavation was carried out in relation to stabilising one of the huts at Cape Adare in 1990 (Harrowfield 1991).

In 2004–05, the archaeological recording, removal and conservation treatment of stacks of provisions cases behind the east wall of Shackleton's *Nimrod* Hut at Cape Royds commenced, the stacks representing both original stores placement and that resulting from the 1960s hut 'clean-up' (Ritchie 2005, 2006) (Figure 2.6). Further work was done in 2007 to remove part of the wall of Bower's Annex at Cape Evans Hut, and further excavate the stables, with the recovery of many artefacts (Ritchie 2008). In the same year two wooden crates were found in ice beneath the Cape Royds Hut, labelled 'MacKinlay's Rare Old Scottish Whisky'. In 2010, these were excavated, and further three cases were found, totalling three cases of whisky and two cases of brandy from Shackleton's expedition in 1908. The bottles in each case were intact and wrapped in straw (Skinner 2014: 73; see https://nzaht.org/category/conserve/shackletons-hut/). The whisky blend was recreated and sold by the original company, garnering substantial publicity for the AHT conservation program (Peat 2013; *AHT News* 01/31/2010).

In the course of this work over 20 years, Ritchie and his associates developed and tested a range of ice-excavation techniques, the most useful of which (given particular conditions) were the use of electric percussion hammers (Ramset Dynadrill), air heaters with directional nozzles, wet-dry vacuum cleaners,

FIGURE 2.6 Stores next to wall at Shackleton's Cape Royds Hut. Eroding cases of tins, stacked both by Shackleton's party and in subsequent 'clean-ups'. Since removed by archaeologists and conserved (Photo by Michael Pearson 1992).

purpose-made steel prise bars and wedges, chain saws for cutting accumulated ice and wrapping in black plastic, especially for thawing artefacts (Ritchie 1990b; 2005). The spreading of local black gravel (scoria) has been used more recently to speed-up seasonal snow and ice thawing over larger areas. A mechanical snow-groomer is now used to clear large annual snowdrifts around the hut, and a 'glycol boiler' heat exchange system has been developed to melt ice (Skinner 2014: 58–59).

The archaeological work in the Ross Sea region has been approached as a tool in the understanding and conservation of the huts. It enabled the artefacts that survived the 1960s 'clean-up' and restoration program to be located, studied and removed where necessary under archaeological conditions. It was targeted at both informing the conservation program, and facilitating it by the controlled removal of artefacts, ice and soil required to allow structural conservation and drainage works to proceed. An archaeological program of survey and analysis of the scatter of objects around the huts, such as was started at the largely undisturbed Mawson's Huts site (see below), was not feasible at the Ross Sea huts, given the extent of previous clean-up and restoration activities. While being targeted at providing information, and retrieving artefacts likely to be disturbed by conservation works, the Ross Sea archaeological research has also produced a cumulative record of the activities and material culture of the 'heroic age' exploration parties, that forms a research and reference collection of immense value for site interpretation and future studies. This has been backed up with detailed recording of the buildings and sites (Gibb et al. 2011).

The Ross Sea Heritage Restoration Project, which has guided the conservation work of the AHT since 2002, has involved a commitment to high-quality materials conservation of the artefacts at the Ross Sea huts, with a conservation laboratory being established at the New Zealand research station at Scott Base near McMurdo for year-round work, and involving more than 60 heritage professionals from around the world (ATCM XXXVI 2013, Background Paper 1 'Antarctic Heritage Trust Update 2013'; AHT website). In 2017, the AHT reached a major phase of the project, in completing the conservation of four historic huts and their artefacts, these being:

- Sir Ernest Shackleton's 1908 base and its collection of more than 6,000 artefacts
- Captain Robert Falcon Scott's last expedition base at Cape Evans and its 11,500 artefacts
- Scott's first *Discovery* expedition base at Hut Point and 500 artefacts
- Hillary's Hut at Scott Base and its 600 artefacts

From 2017, the program refocused on planning for conservation work at Borchgrevink's Hut at Cape Adare, while continuing essential ongoing monitoring and maintenance work at the four Ross Island huts. The finding of an original watercolour by Dr. Edward Wilson, one of Scott's party on its ill-fated 1912 journey to the pole, and a well-preserved Huntley and Palmer tinned fruit cake in the hut in 2016–17 provided further publicity for the AHT conservation program at Borchgrevink's Hut (*History.com* 06/14/2017; *Archaeology* 06/14/ 2017; *Sci-News* 08/10/2017).

With the last major archaeological works at Cape Evans being in the mid-1990s, apart from the annex removal in 2007, and work at Cape Royds in 2004–06, the AHT project has focused primarily on the buildings and materials conservation objectives. There has been no major archaeological component within the conservation program since 2008, such work being limited to occasional artefact removal from deposits, ice and snow removal, and digging trenches to insert water barriers (see Skinner 2014). The planning for the proposed conservation work at Cape Adare has recognised the need for an archaeological component, aimed as a support for the building conservation (Lizzie Meek pers. comm. 27/2/2019), but no broader non-conservation research objectives have as yet been identified.

2.1.2 Commonwealth Bay, King George V Land

The Australasian Antarctic Expedition (AAE), led by geologist Douglas Mawson, established its base at Cape Dennison, Commonwealth Bay, in January 1912, and occupied the hut they built there until relieved in December 1913. The objectives of the expedition were to carry out geographical exploration and scientific investigations in the as-yet unexplored sector of Antarctica due south of Australia and to attempt to reach the South Magnetic Pole. Mawson together with his mentor TW Edgeworth David and Alistair Mackay had searched for and thought they had reached the approximate position of the pole during Shackleton's expedition in 1909, and now another team was to approach it from the opposite direction (Mawer 2006: 175–204).

Mawson's Huts Historic Site consists of four huts:

- a main hut consisting of two conjoined expedition huts, one of which, intended as a base further along the coast, was never deployed and was consequently added to the main living hut as a workshop;
- a magnetometer hut;
- an absolute magnetic hut; and
- a transit hut for astronomical observation.

A scatter of artefacts and other material remains from the expedition, very largely undisturbed by human interference, covers with varying density the entire Cape Denison ice-free area. After abandonment the huts were very seldom visited and had filled with snow by the 1960s (Figure 2.7). Four Australian National Antarctic Expeditions (ANARE) parties between 1974 and 1981 undertook extensive

FIGURE 2.7 Mawson's Hut, Commonwealth Bay. Excavating through snow and ice inside the hut. The ruler sits on the floor boards with a layer of black ice marking the occupation period, with blown snow above (Photo by Steve Bunton 1986).

excavation of snow within the workshop section of the main hut and carried out repairs to the roof (Ledingham 1979). Between 1981 and 1985, three private expeditions re-opened the snow tunnel dug earlier into the main hut.

The first professional archaeological work on huts in Antarctica was carried out on the 1984–85 expedition, by Estelle Lazer (1985), followed by further work in 1985–86 by Angela McGowan (1987), both with the private 'Project Blizzard' expedition. Because Mawson's Huts had experienced such limited visitation and there had been no general 'clean-up' as had occurred at the Ross Sea huts, it retained *in situ* a large amount of scattered artefacts around and down-wind of the living hut (Figure 2.8). These included left-over building materials and packaging, remnants of the masts and wires of the first radio transmitter in Antarctica, survey marks and sample areas for geographical, geological and magnetic survey, discarded equipment and clothing, food containers and other domestic discards. Other material remains were spread less densely across the entire cape. It included the ice-buried frame of a Vickers R.E.P. monoplane, the first airplane intended to be flown in Antarctica which after a crash before departure from Australia was reduced to a wingless air-tractor that saw limited use at the base (Henderson 2009, 2010).

Inside the main hut and workshop was a collection of scientific, domestic and personal artefacts left behind when the hut was abandoned. The archaeological work in the first seasons concentrated on:

- recording and cataloguing the huts and the artefacts inside them;
- commencing the survey of the artefacts outside the huts and defining the extent of the site;
- placing 27 numbered modern objects outside the huts to test the movement of objects due to wind and snow – monitoring continued by Pearson (1986), though only six objects could be found due to snow cover;
- monitoring works and excavating through ice inside the main hut where necessary to provide footings for propping the structure; and commencing testing of the treatments and coating on test panels of timber and metals (McGowan 1988; Hayman et al. 1998).

Materials conservator Janet Hughes was part of the team overseeing artefact conservation and related research on the pattern and rate of material deterioration at the site. The excavation techniques employed by McGowan were experimental, and identified what worked and what didn't in different snow and ice conditions (see Hayman et al. 1998). This test excavation set the scene for Neville Ritchie's more extensive excavations at the Ross Sea huts, and his experimentation with different excavation tools over the following decades referred to above. Subsequent Mawson's Huts Foundation (MHF) reports from the late 1990s, however, do not indicate that Ritchie's reporting of his advances in excavation techniques (e.g. Ritchie 1990b) had greatly influenced planning. At Mawson's Huts, chain saws and variable-speed angle grinders were seen as the best tools, but in the 1997–98 season, the grinders were found to be more difficult to use

FIGURE 2.8 Plume of discarded artefacts downwind of Mawson's Hut (Photo by Michael Pearson 1986).

and were not used in subsequent excavations of ice (pers. comm. Estelle Lazer 24/08/20).

Angie McGowan outlined some of the information that it was hoped would be gathered through controlled archaeological research at exploration huts. Firstly,

> [the] writers of first-hand accounts frequently emphasise those things they see as important and play down or omit other things which they think

> obvious, mundane or derogatory. Their less conscious actions are rarely recorded, as the writers are often too close to the action to perceive the cultural processes in which they participate. Archaeological findings may add extra perspective to the interpretation of an historic site.
>
> *(McGowan 1988: 101–02)*

Further, the ship's manifests and documents requesting stores for donation or purchase provide considerable detail about what stores and equipment were taken south. The distribution of objects across the site tells us what was left behind or discarded, giving an understanding of what was used by expeditioners, and their disposal practices can indicate how they perceived the environment and their place in it (McGowan 1988: 102). The main hut and the artefact scatter are particularly important in this regard (Figure 2.9). In addition, archaeology can augment, corroborate or correct written evidence, fill in gaps in the historical record, and provide new historical and architectural information of use in the conservation process. For example, the documentary record for Mawson's Huts says that no iron nails were used within six feet of the instruments in the magnetograph hut, and only copper nails were used in the construction of the absolute magnetic hut to minimise extraneous influences on readings, yet an examination

FIGURE 2.9 A snapshot of the artefact scatter at Mawson's Huts, showing the diversity of materials discarded from an expedition hut: boots, bamboo poles, heavy trousers, barrel head and the ubiquitous timber from packing cases and construction (Photo by Michael Pearson 1986).

of the fabric of both huts revealed extensive use of iron nails were the written record said there were none (Webb 1925: 75; Lazer 1985: 4–5).

In excavating ice inside the hut, McGowan found staining and physical inclusions in the ice layers that she named 'ice artefacts' (McGowan 1987: 52). Neville Ritchie found similar stain evidence in the Cape Evans Hut in the Ross Sea, but thought 'ice artefacts' a misnomer, and preferred the term 'frozen residues' for the contents of leaking tins or broken bottles, decomposing seal blubber, penguin or husky carcasses, and other animal meat residues found within the ice (Ritchie 1990a: 263).

The next archaeological work at Mawson's Huts was not until a decade after Project Blizzard and was undertaken by the new Mawson's Huts Foundation (MHF) in 1997–98. The MHF was established in 1996 by a newspaper consortium (AAP) as a not-for-profit charity to promote the conservation of Mawson's Huts. The MHF has worked in close liaison with and under the auspices of the Australian Antarctic Division on 14 expeditions to the huts (to 2020), and has cleared snow and ice, stabilised the structure and over-clad the roof to reduce snow infiltration and structural failure. Six of the expeditions have involved archaeologists (Estelle Lazer three times, Anne McConnell, Jody Steel and Stirling Smith) with work involving ice removal, documentation and conservation of artefacts (with professional conservators including Ian Godfrey also involved) and building conservation advice. The perceived positioning of archaeology in the project is reflected in the works plan for the 1997–98 season: 'The first priority of the archaeologist will always be the immediate requirements of the conservation team as identified by the conservation architect' (AAP MHF 1997). At the end of that season, the team leader reported that: 'It was always clear that the conservation program was our primary task and the archaeological program required resources other than those allocated' (Grant 1998). This balance of priorities has remained through the years, and indeed archaeology has been viewed with certain hostility, in part because it competed for scarce resources needed for the 'important' work of physical conservation, and in part because it was perceived as potentially threatening to slow down or modify that work.

One investigative technique tested in 1997–98 was to take ice cores through the accumulated ice inside the main hut so as to observe the stratigraphy of the ice deposition and understand the pattern of snow and ice deposition inside the hut (Hayman et al. 1998). Four ice cores were taken, each about 2 m long, and in addition to their role in illuminating ice stratigraphy, they were returned to Australia for isotope and structural analysis to provide evidence of the physical environment within the hut that impacts on material conservation. Additional snow and ice samples were collected to augment this material in 2002–03. Isotope analysis of the core and snow samples identified seasonal cycles in snow accumulation within the hut, and indicated that there was no significant melt and refreeze occurring. Slow accumulation appears to have taken place in the main hut area from the time of the hut's abandonment in 1914, but the last several metres of accumulation occurred in less than a year, possibly due to the failure of roof lights. Within the workshop hut area, which was cleared of snow in 1978, subsequent snow accumulation was rapid, probably due to the disturbance of weatherproofing during the 1978 work (Donaghue et al. 2014).

The clearing of ice from shelves and other areas inside the main hut in 1997–98 recovered a range of artefacts. An interesting observation was that perhaps the most significant information provided by the artefacts was what they revealed about individual members of the expedition, such as the in situ metal badge from Canterbury College, Cambridge, attached to Eric Webb's bunk, and the reproduction prints, including Fragonard's 'Girl on a Swing' on the wall above the bunk in Mawson's cubicle: 'these artefacts arguably provide a human dimension to the individual expeditioners who are generally portrayed in the literature as remote heroes' (Hayman et al. 1998: 22).

As a result of the tight focus on the conservation of the huts, the only archaeological project carried out by Project Blizzard and the MHF not related to ice clearance and monitoring for building conservation was the survey and documentation of the extensive artefact scatters across the cape, and particularly down-wind of the main hut.

Estelle Lazer noted the potential for special studies at the site during her first visit:

> For example, the distribution of a substantial amount of refuse, especially to the north of the Main Hut, is much closer to the building than might normally be expected, which appears to be a behavioural response dictated by the extreme environment. Such studies provide a different class of evidence to that provided by the historical record and often can yield information about aspects of life that would never be mentioned in the written resources.
>
> *(Lazer 1985: 4)*

The experimental monitoring work commenced in 1985, with the placing of numbered and surveyed objects at strategic points through the main artefacts scatter outside the main hut, to be monitored over a year or longer. This was intended to 'help isolate the variables that may have caused the present distribution patterns of objects on the site. Without some knowledge of how objects arrived at their present locations, a study of spatial behaviour cannot be effectively undertaken' (Lazer 1985: 4). Wind was an obvious causal factor in random artefact distribution, but Lazer noted that 'the distribution of artefacts outside the main hut into definite homogenous groups, such as tin cans and glassware, suggests that their positions can be attributed more to human activity than climatic conditions'. Other variables could be disturbance by the movement of seals and penguins and the activities of later human visitors.

The recording of the artefacts outside the hut was complicated by the short duration of each conservation expedition, and the highly variable snow cover across the site. The recording project was a multi-year undertaking, fitted in around the building conservation work schedule. The survey work done in the 1997–98 season helped define the boundaries of the historic site – essentially the whole of Cape Denison – and gave some indication of the activities and behaviour of individuals during the expedition. It also demonstrated the highly variable seasonal snow cover and ablation rates across the site, with previously exposed artefacts being hidden, while new ones were exposed (Hayman et al. 1998: 22). In 2000–01, after a successful survey during a particularly snow-free season, Estelle Lazer noted that the outcome of three seasons of records by photographs, video recording and filed notes carried out to date needed to be combined into a data base, and backed up by a GPS survey, to make it available and useful (AAP MHF 2001). This work was partially completed by Lazer and subsequently added to as a GIS and image-linked Antarctic Artefacts Database at the Australian Antarctic Division. Together with listing of the artefacts also held by the AAD, it is known as the Antarctic Heritage Register but unfortunately has not been completed due to lack of funding (https://data.aad.gov.au/aadc/artefacts).

The 2000–01 survey recorded previously unseen material such as a sledge, caches of whole and butchered seal carcasses and mooring lines from the expedition ship:

> The results of the outdoor survey confirm previous observations that Mawson's party left material evidence of their occupation right across Cape Denison and onto the ice plateau. It is also apparent that the artefacts across the site can be classified into deliberately and accidentally discarded objects and artefacts that were consciously cached or placed across the site for practical purposes.
>
> *(AAP Mawson's Huts Foundation 2001: 44)*

It had been already observed by Lazer as noted above that the distribution of materials in the scatter closest to the main hut was not random, but to a degree differentiated by material types – it was hypothesised that the this might have been to simplify scavenging for materials for repairs and new uses in an environment where wood, for example, was not otherwise available (Hayman et al. 1998: 23). Similar sorting of material types in dumps of domestic and building materials is found at outback sheep and cattle stations in Australia, similarly isolated from easily accessible sources of needed materials – the dumps are sometimes referred to by pastoralists as 'department stores' (Pearson and Lennon 2010).

Away from the main hut valley many of the objects found were survey markers associated with various scientific and ground survey activities of the expedition (Figures 2.10 and 2.11), and seal carcasses partly butchered, assumed to be caches for emergency use. An unusual find was a dog carcass, located at a staging campsite on the ice plateau to the south of the main base (known to the Mawson expedition as Aladdin's Cave). This was brought down to the main hut, autopsied and found to be an older dog with clear evidence of hard work in its robust muscle attachments and worn teeth, and it had a partially healed fracture of the leg.

FIGURE 2.10 Survey lantern made out of a used tin can by magnetician Eric Webb on the Mawson expedition. The 'make-do' nature of much of the science would not be known of without the material evidence (Photo by Michael Pearson 1986).

FIGURE 2.11 Mawson's Huts site: (a) Transit hut used to house a transit theodolite to take astronomical observations. It shows the severe impact of wind-blown ice crystals – the hut has deteriorated much more in the 35 years since this photograph was taken (Photo by Michael Pearson 1986); (b) Writing on the timber post for the transit telescope, recording the calculated longitude of the site – a valuable baseline location for future research (Photo by Michael Pearson 1986).

Comparisons with journals of the time suggest that it might have been the dog 'Grandmother', a male dog recorded as having died at Aladdin's Cave on August 17, 1912 (Hayman et al. 1998: 26).

In the following season, 2002, conservator Ian Godfrey (in the absence of an archaeologist) continued recording artefacts in an exceptionally snow-free season and also removed what post-Mawson material could be found (Godfrey 2002). Interesting Mawson-era finds included a sledge, a box of finnesko boots, a collapsible boat and paddle, and assorted clothes, shoes and boots, many having been repaired. Further survey work was undertaken by archaeologist Anne McConnell in 2007–08, to the extent that remaining snow cover and avoiding penguin rookeries allowed (MHF 2008). McConnell considered that, apart from the survey of areas still covered in snow and not yet surveyed, the 2007–08 survey (combined with the earlier surveys by Lazer and McGowan) provides some certainty about the distribution of artefacts across the site, and recommended that further surveys should be planned in five- to ten-year cycles to record changes (MHF 2008: 40). Further detailed grid test surveys in the most densely accumulated scatters were carried out near the main hut by archaeologist Jody Steele in 2009–10 (MHF 2010). This detailed grid survey was repeated in 2010–11 by archaeologist Stirling Smith, and it was found that within one year small objects had moved, and some had disappeared from the grid squares altogether, whether by wind, animal movement or visitor interference was not clear, but wind seemed likely to be the main cause (MHF 2011).

Other archaeology continued to be focused on the excavation of ice from areas of the hut such as bunks and kitchen areas, in close collaboration with materials conservators, or in documenting and removing artefact scatters from areas where building conservation works were planned (such as over-cladding of the walls). Smith recommended that, since the principle conservation works in the main hut had now been completed, an archaeological research design should be developed, particularly in relation to the study of the artefact scatters (MHF 2011: 76). This recommendation has not yet been acted upon.

2.1.3 Snow Hill Island, Weddell Sea

Snow Hill Island, in the north-western section of the Weddell Sea, and part of the archipelago off the eastern side of the Antarctic Peninsula, was the expedition base of the Swedish Antarctic Expedition led by Otto Nordenskjöld from 1901 to 1903. The expedition ship, the *Antarctic*, was sunk in 1903 during its voyage to relieve the party, resulting in the stranding for the following winter of the Snow Hill Island party in their hut (Figure 2.12), another three members of the expedition at Hope Bay at the tip of the Antarctic Peninsula, and the rest of the ship's crew on Paulet Island well to the north of Snow Hill. Both the Hope Bay and Paulet Island parties survived the winter in shelters built of local rock, and all three parties were amazingly reunited and rescued by the Argentine naval corvette ARA *Uruguay* in the summer of 1903, with the death of only one man during the winter. The three sites have been declared HSM within the

FIGURE 2.12 Nordenskjöld's Snow Hill Island Hut, 1901–03. The moraine ridge on which the hut sits is unstable, hence the stabilisation works visible here and is being studied for possible climate change induced permafrost degradation (Photo by Michael Pearson 2017).

Antarctic Treaty, and Snow Hill Island Hut was also declared Historical National Monument (Argentina) by the Executive Power Decree # 6058 in 1965 (Capdevila n.d.).

Conservation work on the Snow Hill Island Hut was begun in 1979–80 by a team led by Santiago M. Comerci (lawyer and historian) and has continued with breaks until the present (Figure 2.13), by combined Argentine and Swedish parties, made up variously of Ricardo Capdevila (a lawyer turned historian/museologist – Curator of the Antarctic Museum. National Antarctic Directorate, Argentine Antarctic Institute), Fred Goldberg (a Swedish welding technologist and public climate-change sceptic), JM Ageitos (Argentine museologist) and others. The removal of artefacts embedded in ice was undertaken by Comerci, the ice being melted with screened gas heaters, then exposed to the sun wrapped in black plastic. The as-found position of artefacts was recorded (Comerci 1983, 1981; Capdevila and Comerci 1986; Capdevila 1990; Capdevila et al. 2000).

No archaeologist was involved in the conservation work until 2008–10, after the snow and ice removal and the disturbance of site for stabilisation works had been completed. The archaeologist Victoria Nuviala, employed by the Argentine National Antarctic program, was then mainly involved in cataloguing and field conservation of the artefacts that remained in the hut (Senatore 2019; Nuviala pers. comm. 2019). There is, therefore, no archaeological record of the disturbance associated with the conservation works – which included the removal of up to 1.8 m of ice, the extensive removal and replacement of external and internal felt and tarred paper cladding, and the excavation of trenches around the hut to insert concrete beams for soil stabilisation and as anchors for cables to hold down the hut (Goldberg et al. 2001). The hut sits on top of an unstable moraine ridge of soft clays and silts, possibly effected by permafrost melting, and is at severe risk of subsidence. The soil stabilisation work, commenced in 1996, is still to

FIGURE 2.13 Nordenskjöld's Snow Hill Island Hut: (a) Structure after conservation (Photo by Michael Pearson 2017); (b) Interior of the living area (Photo by Michael Pearson 2017).

be completed (personal observation by Pearson, 2017). Removed artefacts were conserved at the La Plata Museum and then seem to have been sent to the privately owned Ushuaia Maritime Museum for display. A joint Argentine-Swedish archaeological party commenced a new program in 2020, led by Dr. Pablo Fontana, Argentine Antarctic Institute and Professor Dag Avango, Luleå University of Technology. Work included photographic and digital documentation of the Snow Hill Island Hut and the stone shelter at Hope Bay. In the case of the Snow Hill Island Hut, the digital modelling is to provide a base record to support work on understanding and hopefully remediating instability/permafrost degradation issues, which are climate change impacts (Swedish Polar Research Secretariat 2019; Almevik et al. 2021; see https://meltinghistory.org).

The stone-built Hope Bay shelter (Figure 2.14), near the Argentine base Esperanza, was described in earlier reports as having partially collapsed and being rebuilt in 1992–94 from the remains and with the addition of other materials from the nearby abandoned British base. There was no professional archaeological input, with artefacts from the base of the hut being removed for display in Ushuaia. The digital recording undertaken in 2020 might indicate the extent of survival of the original construction. The Paulet Island shelter (Figure 2.15), a much larger and more substantial structure, was better preserved, though again the clean-up and stabilisation work was undertaken with no controlled archaeological recording or oversight (Capdevila nd; Capdevila and Ageitos 1993; Capdevila et al. 1994).

FIGURE 2.14 The shelter at Hope Bay, where Nordenskjöld's party members Anderson, Duse and Grunden wintered in 1903. Argentina's Esperanza Base behind (Photo by Michael Pearson 2017).

FIGURE 2.15 Shelter in which the shipwrecked crew of Nordenskjöld's ship Antarctic survived the winter of 1903 on Paulet Island (Photo by Michael Pearson 2008).

The Hope Bay and Paulet Island shelter sites have much in common with shipwreck survivor campsites, which raises potential avenues of research that are discussed in the analysis at the end of the chapter.

2.2 The archaeology of late-nineteenth and twentieth-century scientific stations and field camps

A number of mid- to late-nineteenth century expeditions that preceded the 'heroic age' were driven by two scientific objectives: the transit of Venus and the documentation and study of the earth's magnetic field, and were either ship-based or, as we will see, were located on the sub-Antarctic islands. The scientific objectives of the 'heroic era' tended to be more diverse, reflecting the longer periods of time spent at one location (the huts) and the associated opportunities to undertake broader science. Except in one case, Bruce's weather station in the South Orkneys which became the permanent Orcadas Station, the 'heroic era' scientific work was short-lived, and serious national attention only began to be paid to establishing scientific bases in Antarctica during the Second World War. International tensions and overt interest in establishing territorial claims over Antarctica persisted during and after the war and were intensified by Cold War rivalries. The bases established at this time, while clearly making a territorial statement, were also seen as having a scientific rationale. The scientific aims overwhelmed the territorial claims during the International Geographical Year

(IGY) programs in Antarctica, culminating with the agreement of the Antarctic Treaty in 1959 and its ratification in 1961. There are a number of stations established in this era that have had some degree of archaeological survey, and this work will be described in the relevant geographical areas outlined below.

2.2.1 South Orkneys

Discovered in 1821 by sealers George Powell and Nathaniel Palmer, Laurie Island in the South Orkney archipelago was chosen by William Spears Bruce, leader of the Scottish National Antarctic Expedition, as the location for his scientific weather station at Scotia Bay in 1903. Bruce built the weather station and refuge named 'Omond House', together with a wooden magnetic hut named the 'Copeland Observatory'. The weather station has been continuously operated by Argentina since 1904, when Bruce departed, and renamed Orcadas Base, the oldest continually occupied base in Antarctica. A prefabricated timber building was brought to the site in 1905, now named Moneta House, which provided living quarters, Bruce's party having occupied their ship *Scotia* or Omond House during their stay.

Omond House (Figure 2.16), built to act both as the base for meteorological observations, as an accommodation when the *Scotia* was undertaking oceanographic cruises and as a refuge in case of loss of the ship, had four hammocks

FIGURE 2.16 Omond House, South Orkney islands, before its reconstruction (Photo by Claudio Parica ca. 1985).

and two beds with a cooking area and meteorological 'office'. It was a squat stone-walled building roofed with timber trusses and canvas covering, and over 100 tons of rock were used in its construction. The ruined remains were excavated in 2000–05 by archaeologist Veronica Aldazabal, who exposed the original timber panelled floor within the collapsed walls, recovered artefacts from the expedition and determined subsequent uses of the building. The artefacts included food can lids, cardboard packaging for dried vegetables and condensed milk without sugar, together with local resources, especially penguin eggs and meat. Fragments of glass, wine bottles, stoneware bottles, jars and containers sealed with cork and wood bungs were also found. Scientific samples, including penguin remains complete with skin and plumage, bird carcasses and eggs with embryos were identified. The excavation and the material remains recovered allowed spatial interpretation of daily activities and the living conditions of both the members of the Scottish expedition and those of the first group of Argentine winterers (Aldazabal 2007).

A study of the fabrics, rope and leather identified containers such as storage bags for coal, vegetables and personal items; canvas that may be roof and wall covering, sometimes nailed with tacks; curtain materials probably used to divide the hut spaces; and cotton and woollen clothing (Aldazabal 2012, 2017). Moneta House, restored by Ricardo Capdevila apparently without archaeological input, was converted into a museum of the Argentine weather station occupation as part of the project.

2.2.2 Kerguelen

The scientific observatory site at Observatory Bay on Kerguelen Island was surveyed and excavated in 2006–07 (Courbon and Le Mouël 2009) (Figure 2.17). Established in 1874 as one of two stations erected by a British scientific party under L.J. Wharton on board HMS *Volage*, to observe the transit of Venus, the observatory site was subsequently occupied by a German scientific party under Erich von Drygalski in 1902–03, to study the terrestrial magnetism (Lüdecke 2004), then the area was occupied by an elephant sealing party (the two try-works of which survive) and a French party testing sheep breeding on the island. Three main structures at the site were excavated, most evidence relating to the 1902–03 German phase, but also including the foundations of the 1874 British main building beneath the German dwelling house. The main buildings were identified as the dwelling house (Wohnhaus), an absolute magnetism observatory (Absoluthaus) and an observatory of relative magnetism (Relativhaus). Collapsed timber wall and roof structures were found, and were the basis for on-paper reconstructions of the original buildings. The archaeological evidence of some of the buildings differed from the archived plans and documents of the expeditions. In the case of the Relativhaus, for example, it was designed to have a suspended timber floor, but archaeological evidence shows that the instruments were in fact installed on the ground. It appears likely that at the time of construction it was

FIGURE 2.17 Astronomical observatory, Observatory Bay, Kerguelen Island. Excavated in 2006–07 (Photo by Paul Courbon 2009).

perceived that a wooden house with crawl space beneath would not be stable given the strong winds of the area, so the idea of such a floor prone to vibration was abandoned. As at Mawson's Huts, the archaeological evidence can inform the as-built scientific infrastructure, modified due to practical on-site conditions, as opposed to what was planned on paper. Le Mouël also surveyed a landing place and a surviving brick plinth thought to be for astronomical instruments for the British transit of Venus party at Supply Bay (Le Mouël 1996: 44–51).

Scientific stations were also established on Kerguelen by a German party at Betsey Cove to observe the transit of Venus in 1874, being surveyed by Le Mouël in 1995 (Le Mouël 1996: 67–74). Another station was set up by a US party at Point Molloy, while other parties went to Campbell Island and Auckland Island south of New Zealand (see below), and to Ile Saint Paul north of Kerguelen. Only one of these stations has been investigated archaeologically.

2.2.3 Auckland Islands

In 1840, three expeditions made use of Terror Cove, Port Ross, on Auckland Island, south of New Zealand, for magnetic observations – including those led by Captain James Clarke Ross in HMS *Erebus* and HMS *Terror*; Lieutenant-Commander Cadwalader Ringgold, of Charles Wilkes's US Exploring Expedition, in USS *Porpoise*; and Captain Jules Dumont D'Urville in *L'Astrolabe* and *La Zélée*. Ross and D'Urville seem to have used the same site for their observations, but because the process required minimal land-based infrastructure, no archaeological remains of any of the sites have been found, though it has been suggested that some of the remains associated with the 1874 observatory (see below) may have been from Ross's (Jones 2009). While Ross was the last of the expeditions to visit the bay, because of the predominance of British maps, his eminence in the scientific field and the fact it was Territory administered by New Zealand, a British colony, it is his and his ships' names that have been adopted for the geographical features.

A German station to observe the transit of Venus was established at Terror Cove in 1874 by Lieutenant's Becks and Siegel in the hired ship *Alexandrine*. Their work was coordinated with that on Kerguelen described above. The combination of photographic records, drawn plans and physical evidence has enabled the layout and features of the observatory complex, which apart from the instrument bases and shelters, included a large accommodation house and two separate huts, to be defined. Of the five brick instrument pedestals that have been identified, one has been suggested as being Ross's, but site was also reused by two other expeditions in 1907 and the 1940s, so attribution is not certain (Jones 2009).

2.2.4 South Georgia

The remains of the German Science Station, organised as part of the First International Polar Year, established at Köppen Point in Little Moltke Harbour, on South Georgia, in 1882–83, were recorded by the South Georgia Archaeological Project in 2019. Established to study the transit of Venus, meteorology, geomagnetism, tidal patterns, geology and wildlife, the remains of the station consist of timber beams, stone foundations and wooden floors, and the circular iron remains of the observatory (Neil 2019).

2.2.5 Stonington Island

East Base, the oldest remaining US research station in Antarctica, is located on Stonington Island on the west coast of the Antarctic Peninsula, and was occupied by a US Antarctic Services Expedition (1940–41) and subsequently by the Ronne Antarctic Research Expedition (1947–48) (Figure 2.18). The initial establishment of East Base was in response to active pursuit of Antarctic sovereignty claims by Germany and Argentina in the lead up to the Second World War. Two American bases would be established, East Base, and West Base ('Little America III') near the Bay of Whales, which was built on an ice shelf and eventually calved into the Ross Sea (see below). East Base was to support aerial surveys of the Antarctic Peninsula region. Britain's Falkland Islands Dependency Survey (FIDS) established their 'Base E' 100 m from the US East Base in 1946, and operated it until 1975, while also utilising part of the East Base buildings. The East Base buildings were vandalised by Chilean and Argentine visiting ships before Ronne arrived in 1947, and the base was declared an HSM in 1989 (Broadbent and Rose 2002: 238–42).

East Base consisted of four buildings (science building, main building, machine shop and storage shed) and housed 26 men for the winter of 1940,

FIGURE 2.18 East Base, Stonington Island, US station 1940–48. Archaeological investigation took place in 1991 by Robert Spude and Catherine Blee and in 1992 by Noel Broadbent. (Source: http://picasaweb.google.com/geoffrey.boys/2007OurFirstPicturesOfAntarcticaInFebruary#5038576672732785698).

before evacuation due to the start of the war. A 12 by 12 foot (3.6 × 3.6 m) outpost building was added near the main building, and this became the base for Finn Ronne's Expedition of 1947–48, which saw the first overwintering of women (Edith Ronne and Jennie Darlington) in Antarctica.

In 1989, Peter J. Anderson of the Byrd Polar Research Centre at Ohio State University and Karen Tupek, daughter of Finn Ronne, proposed a historic preservation research project, encompassing East Base, Palmer Station and sealing camps on the South Shetland Islands to be administered by the US National Science Foundation (NSF), but no appropriate program or funding was found (Broadbent and Rose 2002: 250). An East Base clean-up project, under the direction of NSF personnel, was carried out as a part of extensive environmental clean-up operations in 1991–92, coinciding with the adoption of the Environmental Protocol to the Antarctic Treaty. The clean-up also enabled the site to be preserved and interpreted as a monument to the legacy Richard E. Byrd (Broadbent, Blee and Spude 1992; Broadbent and Rose 2002: 238).

In 1991, archaeologist Catherine H. Blee (later Spude) and historian Robert Spude undertook a four-day project at East Base, which included a preliminary survey, inventory and the selected removal of artefacts (Spude and Spude 1993).

The East Base Project was the first US effort in historical archaeology in Antarctica (Broadbent 2009: 55). The resulting report for the US National Parks Service recommended collection, storage and curation of the artefacts of the site; ongoing monitoring of the condition of the site; and covering surface artefacts and hazardous materials with a layer of gravel to both protect the deposits and present a more attractive site for visitors, and to preserve them for future archaeological investigation, as they

> ...have the potential to yield important information of interest to scientists studying human disposal patterns or the degradation of materials (medicines, organics) over time. Historians, sociologists and archaeologists are increasingly discovering that the difference between what people say they did and what actually occurred can be revealed through archaeological methods and techniques. It is incumbent upon preservationists to protect such outstanding records as those provided by the domestic dumps at East Base as a repository of knowledge.
>
> *(Spude and Spude 1993: 149)*

The Spudes saw both practical and research potential in the last remaining food caches at the site:

> The caches are also significant for the scientific information that may be contained therein. An analysis of the spatial location of different food types in the cache may provide information on how provisions were organized in camp. Other scientific data may include answers to questions on the relative decay of metals in a polar environment, the length of time canned foods remain edible, and the types of foods believed to be of nutritional value in harsh polar conditions in the middle of the 20th century.
>
> *(Spude and Spude 1993: 129)*

Based on the initial report, a team including archaeologist Noel Broadbent and British expeditioners from the British Rothera Station went to the site and carried out the clean-up and urgent repairs over 12 days in 1992. While few original objects remained within the huts, the snow-melt revealed a large scatter of objects among mixed 'trash and wooden debris'. Areas were examined, exposed artefacts 'of historical interest or in danger of further damage' were collected and some hazardous materials were removed.

> The artifacts consisted of a wide range of objects that reflected the daily tasks and concerns of expedition members. Among numerous items of discarded clothing were boots of all kinds, gloves, and mittens. Hand tools, turnbuckles, sled parts, dog harnesses, collars and chains, metal stakes, trail flags, crampons, ice axes, weather balloons, ring markers for birds, and containers for biological specimens suggest the amount of equipment

> required by exploring parties. Meanwhile numerous bandages, razors, magazines, books, playing cards, Monopoly cards, chess pieces, light bulbs, radio tubes, food cans, jars, bottles, plates, cups, cutlery, and many other domestic objects shed light on how the men and women of East Base passed both their working and leisure hours.
>
> *(Broadbent and Rose 2002: 253)*

At the adjacent British 'Base E', the original 1946 hut was destroyed by fire, and the current hut was built in 1961. Although conservation work was carried out at the site starting in 2017 by the United Kingdom Antarctic Heritage Trust (UKAHT) with support from the New Zealand Antarctic Heritage Trust (NZAHT), there has been no archaeological program associated with the work. Similarly, current UKAHT/NZAHT work at 'Base A' at Port Lockroy (1944–62) (initially surveyed, cleaned up and partially excavated in 1994–96 without archaeological input – Burkitt 1996; Cochran 1996); and 'Base Y' on Horseshoe Island, in Marguerite Bay, does not appear to have an archaeological component. In all three cases, there are insufficient available reports to demonstrate that archaeological work is or is not desirable (websites for the UKAHT and NZAHT).

2.2.6 Australian Antarctic Stations, East Antarctica

Wilkes Station was constructed by the United States in February 1957 as part of Operation Deep Freeze II during the IGY Program (Figure 2.19). The US Navy unloaded 11,000 tons of building material, fuel, food and scientific equipment onto Clarke Peninsula. The station was over-supplied, and much of the stores landed was never used and remains on the site. The station, to house 24 men over winter, consisted of a number of 'Clements' buildings constructed with prefabricated insulated panels on concrete foundations, with additional 'Jamesway' insulated canvas tunnel huts used as temporary accommodation and storage. Wilkes was offered to Australia as a base in 1959, and additional huts and connecting corridors were built in the 1960s. The site, however, was prone to snow build-up, and the 'temporary' base was abandoned by Australia in 1969, its operations moving to the nearby Casey Station. The latter was itself abandoned in 1988 and demolished by 1993, the station then being transferred to the current Casey Station nearby.

Because of the ambivalent ownership and environmental responsibilities for Wilkes (being United States owned and Australian occupied), and the substantial problems that would have to be faced in removing the buildings and stores that are now largely ice-anchored, clean-up has been limited to hazardous and wind-blown materials (see Snape et al. 1998). A master plan for plotting and documentation of all the buildings and remains and a preliminary assessment of building construction and condition were undertaken by archaeologist Elspeth Wishart and conservator Linda Clark in 1988–89, but no archaeological program has followed up on this initial work (Clark and Wishart 1990). Wilkes remains the most

FIGURE 2.19 Wilkes Station, established by the United States in 1957, operated by Australia in 1959–69, then abandoned. The most intact of the IGY Antarctic station and a major heritage and environmental dilemma (Photo by Michael Pearson 1986).

intact surviving IGY station, but its study has been extremely limited. While the Wilkes station is undoubtedly of historical and archaeological significance, the scale of waste and contamination at the site has meant that it has never been suggested for HSM listing by either Australia or the United States and remains an embarrassment for both countries given their environmental protection obligations under the Madrid Protocol.

Five other stations were established by Australia in the 1950s: Casey, Davis, and Mawson on the mainland, Atlas Cove on Heard Island, and Macquarie Island Station at Buckles Bay. Each of these has had some degree of heritage recording and assessment, by archaeologists and others (e.g. Clark and Wishart 1991, 1993; Rando and Davies 1996; Staples 1996; Grinbergs 2000; Vincent 2004), but no major archaeological work has been carried out at these sites, despite repeated recommendations in the various heritage reports that such work is needed.

Similarly, the United Kingdom, which had by 1992 abandoned 18 bases and two huts built since 1949, carried out an assessment of each site to determine the need for clean-up of waste and the potential of the sites for conservation as an HSM, resulting in the paper 'Abandoned Stations and Field Huts: The British approach to management' (ATCM XVII, 1992; also Shears and Hall 1993). The absence of appropriate skills and resources was recognised and led to support for the establishment of the UKAHT. The latter then collaborated with experienced conservation staff from the NZAHT to survey the abandoned sites in the

Antarctic Peninsula region commencing in 1994 (ATCM XVIII, 1994, Agenda 4E, 'Clean-up and conservation of abandoned British bases in the Antarctic Peninsula region'), including the sites referred to above. Again, the survey and study work included no archaeological component.

2.2.7 Ross Sea area

The first scientific stations established in the Ross Sea after the 'heroic era' were a series of bases in the Bay of Whales area established by Admiral Richard Byrd in 1929 (Little America I), 1934 (Little America II), 1940–41 (Little America III), 1946–47 (Little America IV) and 1956 (Little America V) (Carter 1979). All of these stations were subsequently buried deep within the ice shelf, or were carried out to sea within icebergs, and as a consequence, none has been studied archaeologically. McMurdo Station, established as a naval air facility in 1956 as part of the US Operation Deep Freeze, is located in McMurdo Sound near the 'heroic era' Scott and Shackleton huts and was subsequently the centre for US activities during the IGY 1957–58. It remains as the largest human settlement in Antarctica, supporting up to 1,200 people. While historically important buildings appear to survive within the modern station (discussed in Davis 2017; Petersen 2021), there has been a constant development and churn of structures, so what remains of heritage or research value is entirely unclear. The two HSM declared at McMurdo are not for buildings or historic remains but are a statue of Richard Byrd, and a plaque commemorating the now-removed nuclear power plant (1962–72). The authors are not aware of any archaeological survey or research at McMurdo Station, though a large amount of cleaning-up started in 1992, coinciding with archaeological work by US parties at Marble Point and East Base. Over recent years plans for the total redesign and reconstruction of the station have been reported in the press, which might indicate the last chance for archaeological survey and research.

A US Naval Construction Battalion Reconnaissance Unit (Seabee) camp was established at Marble Point, 80 km from McMurdo Station in 1957 in preparation for the building of a rock runway for aircraft operations. The intended 10,000-foot (3,048 m) runway was never completed due to its cost and unreliable sea access and is now used as a helicopter refuelling point with related accommodation and support buildings. Archaeologist Noel Broadbent surveyed the Marble Point airfield and construction camp in 1994, but this was after clean-up programs of 1959 and 1989–91, which had removed the original buildings and infrastructure. Broadbent mapped the Marble Point North Base site and collected kitchen and dining items, office supplies, clothing, hut and tent fittings and personal items, reflecting the everyday life of the Seabees and later crews that are seldom reported in the official record (Broadbent 1994). A large piece of physical evidence of the air history of McMurdo is the wreck of a Lockheed Constellation aircraft 'Pegasus' which ran off an ice runway south of McMurdo in 1970 (Figure 2.20).

FIGURE 2.20 The wreck of the 'Pegasus' constellation aircraft, which ran off an ice runway and irreparably damaged in 1970, near Ross Island (Photo by Michael Pearson 1991).

Nearby McMurdo Station is the New Zealand Scott Base, established in 1957 to support the Commonwealth Trans-Antarctic Expedition (TAE). The A Hut built for winter occupation of 23 men of the TAE, and now called Hillary Hut after the leader of the party Sir Edmund Hillary, was restored in 2017 (Watson and Ussher 2018). The hut was moved 40 m in 1989 to distinguish it from the other base buildings. Two small detached magnetic research huts also survive. The conservation project had no archaeological component.

2.3 Discussion of the archaeology of huts and scientific stations

Ritchie and Simmons in 1987 recognised that the material remains of the early expeditions were 'time capsules of technological achievement, vital sources of knowledge about undocumented facets of the expeditions' (Ritchie and Simmons 1987: 1). They also saw the archaeology of the buildings themselves as offering an interpretation of structural history, especially where later changes were made, such as that by Shackleton's Ross Sea Party in 1914–17, that altered the structure of a hut in ways that had no documentation. This evidence, together with the related artefacts, provided an insight into the conditions faced by the marooned Ross Sea Party and their responses to those challenges that were not provided by documentary sources.

The archaeological work in support of the conservation is a valuable and necessary use of archaeology. Archaeology has proven to be an essential tool in building and site conservation at both the Ross Sea huts and Mawson's Huts. The first plan of management for the Mawson's Huts was written in 1993 by an archaeologist (Pearson 1993), and archaeologists have been directly involved with the ongoing conservation planning process resulting in the current iteration of the plan (Australian Antarctic Division 2013). Archaeologists have been central to the development of management plans at other sites such as those in the Ross Sea, and on several of the sub-Antarctic islands. This work has focused on documenting the archaeological resource for management purposes: what has been lacking is research-oriented strategic planning and commitment of resources for non-conservation archaeological investigations. This results from the particular context in which the management plans are developed, both by national land management agencies and within the ATS, where the plans are seen as a means of guiding management and conservation, rather than as tools to encourage or direct research.

The experience of the work of the NZAHT at huts in the Ross Sea, and the MHF at Commonwealth Bay has shown that as conservation work progressed the perceived need for archaeological work declined, while the role of materials conservators working on the artefacts had increased. No archaeologist has worked at the Ross Sea huts since 2008, or at Mawson's Huts since 2011. The challenge for archaeology is to develop and have adopted a detailed research strategy for non-conservation-related archaeological research, with clear objectives and potential outcomes that would attract the scale of funding and logistical support necessary to raise an expedition or to expand an existing conservation-based expedition. Such a strategy needs to consider the research aims that might apply to the different phases of Antarctic history, including the sealing era, the 'heroic era' expedition bases and field scientific sites, the whaling era in the Antarctic Peninsula (and the closely related and more numerous sites in the sub-Antarctic islands), the post–World War II expansion of national programs and the IGY, and the activities undertaken within the ATS (now over 60 years old). The long-term archaeological research of the sealing era sites in the South Shetland Islands has been guided by an ongoing and to some extent opportunistic academic program but has not yet been formalised as a long-term research strategy or plan that can be critiqued and reviewed. The lack of compelling research strategies and the resulting failure to attract research funding support is a recurrent problem for archaeology in Antarctica and the sub-Antarctic.

One of the influences of archaeology has been to broaden the understanding of the wide range of material remains that can constitute valuable evidence of the past in Antarctica, and to encourage managers to acknowledge the potential for 'rubbish' to be seen as 'artefact' and worthy of documentation and study (e.g., Pearson 2004). The archaeological work has shown that what was regarded as rubbish and commonly disposed of in the 1960s clean-ups can in fact have major historical and research significance. It contains evidence of the more prosaic and mundane aspects of survival in the polar environment, the things that do not always make their way into the diaries and official records of the expeditions.

Some examples from Mawson's Hut illustrate the point. Material evidence can show how the occupants of the site made use of what was available to them to solve problems not envisaged when they packed their ships to leave home. Examples would include the lamp, made out of an old tin can, used to mark a reference point needed for magnetic observations during the long winter's night at Mawson's Hut; and the strips of metal, canvas sail and insulation material used to seal the joints between the roof boards of the hut to reduce the penetration of wind-blown snow – a problem created by the combination of extreme winds, snow and the amateur quality of the hut's construction, not envisaged when the huts were designed. Some items that might appear to be rubbish can be something else entirely, such as the jumble of wires and long post sections that are the remains of the aerial set-up to make possible the first radio transmissions from the Antarctic continent. And then there are the carcasses of seals and penguins, killed and stored to provide emergency food supplies, which have the potential to provide chemical analysis of the atmospheric and environmental conditions of 1911 (Hughes 2011).

Spatial analysis of artefact scatters could give clues of the changes to human behaviours, such as rubbish disposal 'shortcuts' taken in the extreme climatic conditions of a site, and the conscious segregation of materials that might prove useful for repairs and use at a later date (as is suggested at Mawson's Hut). Such analysis first requires the study of all the factors that could influence the spread of material such as wind, animals and post-deposition human interference, as well as the conscious and unconscious actions of the expeditioners themselves. At the Ross Sea huts, the study of the stacks of food cans outside the huts, otherwise potentially seen as being 'all the same', could indicate preferences to particular food types during the course of the expedition by what is left over at the end. For example, a personal and unsystematic observation at Cape Evans Hut (by Pearson) suggests a lot of tinned Scottish kale was not eaten, probably referring to a dietary bias shared by the observer.

The potential of 'rubbish' to be valuable as 'artefact', however, is just one factor to be considered in the ongoing tension between the identification and protection of cultural heritage and environmental repair through the removal of human waste and contamination, a tension reflected in the Madrid Protocol's Annexes III and V. This is perhaps most dramatically illustrated in the case of Wilkes Station, as outlined above. What the ATS has not yet achieved is a mechanism that requires an informed dialogue between the cultural heritage and environmental protection needs, and an open decision-making process to agree actions that satisfy, as far as is feasible, the respective objectives.

An emerging consideration in the Antarctic and sub-Antarctic (as it is in the Arctic) is the identification of cultural landscapes, and the role that archaeological sites can play in that context (Barr and Pearson 2021). Studying archaeological evidence within a cultural landscape context stimulates ideas about the relationships between human material evidence and its locality, and the natural environment, widening the scope for interpretation of that evidence and suggesting alternative protection and conservation solutions beyond the limits of a hut or building. Cultural landscapes include not only physical aspects of human and environmental processes but also meanings and values (Denham 2017). While much of the discussion of landscape in Antarctica has focused on cultural impacts as a degradation of wilderness values (e.g. Tin et al. 2008; Summerson 2013), rather than on the values and protection of cultural landscape, some have discussed the tensions in that dichotomy (e.g. Evans 2011).

Cultural landscape has entered the lexicon and the policy thinking of the ATS within the last decade. The clearest expression of the official recognition that Antarctic cultural landscapes exist and have management implications is in relationship to Mawson's Huts. Mawson's Huts group is part of a cultural landscape in which the huts and other archaeological remains are central components. The immediate area of the four timber huts was listed as Historic Site and Monument No 77 in 2004, and an Antarctic Specially Protected Area (ASPA) and an Antarctic Specially Managed Area (ASMA) were created to protect the site. In 2014, the area protected by the ASMA and ASPA were de-designated and expanded to provide additional protection for the cultural landscape at Cape Denison and to simplify the management arrangements for the site. The cultural landscape was identified as having both important material remains such as artefact scatters and food caches (the extent of that scatter as a component of the cultural landscape being defined by the archaeological survey over the previous 20 years) and visual symbolism of the challenges facing humans in its occupation. As the ASPA management plan says,

> Cape Denison is not only the windiest place in Antarctica, but also the windiest place on Earth at sea level. The site thus demonstrates the physical and symbolic context of the extreme isolation and harsh conditions endured by the expedition members and, by association, by all other 'heroic age' researchers and explorers. In designating the entire area as an ASPA, Cape Denison's unique 'sense of place' is protected, with Mawson's Huts and Boat Harbour as the focus of the visual catchment.
>
> *(Australian Antarctic Division 2013; Barr and Pearson 2021)*

Another area with high archaeological and cultural landscape values is the Byers Peninsula on Livingston Island, South Shetlands. There the many sealer's shelters are scattered across the north and south beaches and on the west coast overlooking Plymouth Harbour. The landscape is one of multiple associations – this is the landscape the sealers had to work in and exploit for both the seal resource and for their domestic needs. The presence of the simple stone and whale bone shelters spread widely amidst the stark landscape of rock stacks and headlands, the rugged coast backed by the snow and ice of the interior, and the naming of key points by the sealers, shows this as a human landscape sewn with cultural meaning, not just a wilderness spoiled. The ASPA plan for the Byers Peninsula recognises the presence and significance of the human settlements, and the 2022 edition of the management plan clearly identifies the cultural remains as one of the values of the ASPA, and lists all the known sealing site locations (Management Plan 2022). The next step will be to integrate this knowledge into an understanding of the area as a cultural landscape. This should be considered during its next revision, a suggestion reinforced by the inclusion, in 2018, of 'cultural landscape' as one of the attributes of a site that should be considered in its description and documentation in the *Guidelines for the assessment and management of Heritage in Antarctica*, developed for the Antarctic Treaty Consultative Parties (ATCM XLI, 2018, Resolution 2, Annex).

The rise of consciousness about cultural landscapes is part of a promising trend that might help counter the traditional reliance on documentary evidence alone to define what is of historical importance. If an event or activity in the past is well documented (such as in 'heroic era' journals) it is likely to be recognised as being of historical and heritage importance, while events and activities with little or no documentary evidence are not valued. The result of this bias is seen in the HSM list created within the ATS. An analysis of the HSM sites reflecting the first 100 years of Antarctic history, from 1820 to 1920, found that of the 35 places listed 30 commemorated the 'heroic era' expeditions while only two sites commemorated the extensive whaling history and none at all commemorated the physical remains of the sealing era (Senatore and Zarankin 2014: 126–127; this bias has also been observed in Barr 2018). Sealing has a rich archaeology, but a relatively poorly documented history, and the textual evidence that does exist is usually not specific about the location and dates of events and activities, yet that inadequate historical record has been used in the context of HSM listing as the trigger for commemorative recognition. The 'construction of memory' is seen to be biased towards the textual rather than the material evidence of the past. Furthermore, within the 'heroic era' context, the conservation of the interiors of the huts themselves and the relocation and presentation of surviving artefacts are commonly based on documentary and photographic evidence of the time, rather than the actual condition of the huts or the location where artefacts were first recorded or located archaeologically. This completes a circle in which historical images establish a cherished image of the past, and the physical evidence is altered to recreate those images. As indicated earlier, María Ximena Senatore has called this the 'archaeology of nostalgia' (Senatore 2019).

The Hope Bay and Paulet Island hut sites, and the Shackleton *Endurance* campsite on Elephant Island, have much in common with shipwreck survivor campsites, a class of archaeological sites that has been studied elsewhere over recent decades. Other shipwreck survivor camps are associated with the sealing sites at Heard Island and in the Prince Edward and South Shetland Islands, as discussed below. At shipwreck survivor camps in other parts of the world research questions have been asked that are directly relevant to the Antarctic examples (see Gibbs 2003). Some characteristics of survival camps in more temperate climates do not apply in the Antarctic – such as the need for defence from indigenous inhabitants, the option of escaping overland, and the necessity of setting up lookouts for passing ships (there weren't any) – but others do, such as the challenge of accessing food and materials for survival, and the potential for internal division and dispute among the wrecked group. For example, does the hierarchy of power and related control mechanisms that exists aboard ship automatically transfer to the context of a wreck camp, or does it have to be re-negotiated? In some wreck sites around the world the hierarchy was clearly established by dominant officers, while in others (such as some of the Dutch wrecks on Australia's west coast in the eighteenth and nineteenth centuries), hierarchy broke down because of weak leadership and forms of sometimes bloody anarchy resulted. The historical accounts of the elephant sealer *Trinity* wreck on Heard Island, for which the archaeological sites survive, is a clear example of the problems of maintaining authority and control (Downes 2010). Access to materials for survival was critical – where the wrecked ship was close to land it could be used as a source of building materials and food, but where it was not the local resources had to be called upon. In Antarctica this directly influenced the form of the survival structures built, the ability to be able to construct rescue vessels, and the need to collect food stores during the remaining summer.

The loss of the *Antarctic* during the Nordenskjöld expedition far offshore in the Weddell Sea meant that there was not sufficient timber available to construct an escape vessel at the Paulet Island survivor camp. At several of the South Shetlands wrecks sites, however, escape vessels were being built at survivor camps when rescue arrived in the form of other sealing parties. The option of sending a boat off to arrange a rescue was not a realistic one in most Antarctic circumstances, though Shackleton did so successfully in his epic boat journey from Elephant Island to South Georgia in 1916. In an Antarctic situation the added stress associated with not being able to build your own escape boat, and the extremely unlikely occurrence of a passing ship after the end of the sealing season, must have influenced the minds and behaviours of the wrecked party, and consequently the archaeology of the site (see Gibbs 2002). As Gibbs has noted, 'Studies of the material culture of survivors could consider issues of perceptions of property, ownership, sentiment, and the sharing or management of resources in such crisis or survival situations' (Gibbs 2003: 133–134). It remains to be seen if the archaeological research opportunities at Paulet Island have already been largely lost due to the extensive disturbance caused during clean up and reconstruction works but such potential does appear to survive at some other Antarctic and sub-Antarctic wreck sites that have yet to be studied.

A key opportunity offered by sites of historical scientific observation is the potential to use the same sites for repeat or continuing observations of the same phenomena. Orcadas Base in the South Orkney Islands, established in 1903 and operated continuously since that date, is an outstanding example of continuous data collection. Scientific sites were established to provide baseline observations in such fields as magnetic research, meteorology, astronomy, biology and geology. Opportunistic repeat observations for magnetic and climatic data has occurred at Mawson's Hut and at Auckland Island, and the identification and protection of known observation points at other stations provides the potential for future repeat observations. Other scientific research potentially available at historic sites includes analysis of datable biological remains for correcting Antarctic C14 dating anomalies, and baseline contamination studies for certain chemicals such as mercury (Hughes 2011: final chapter). Broadbent discusses the potential to use archaeological evidence as input into global climate change research in the Antarctic, as the historic sites provide baseline examples of earlier environmental conditions (Broadbent 2009: 59).

As has been seen, the conditions under which scientific data was collected, as reported in the literature, may be found through archaeology to be incorrect. Examples of these anomalies include the observed use of iron fastenings in the magnetic observation buildings at Mawson's Huts, and the shift from suspended floor to on-ground instruments at the Kerguelen observatory. The implications of the observed differences in the interpretation of the results of the relevant scientific programs have not as yet been considered as far as the authors are aware.

2.4 The archaeology of sealing, whaling and penguining in the Antarctic and sub-Antarctic

In Chapter 1, we outlined the range of animal species exploited in the southern oceans, for sealing, whaling and penguining and the focus on two distinct products: the hides and dense underfur of the fur seal and the oil of whales, elephant seals and penguins.

Sealing, as will be seen, required little specialised technology: ship and boats, some basic tools and salt for curing skins, and if elephant oil was sought, try-pots in which to render the blubber into oil. Sealing was pursued on sub-Antarctic islands wherever seals could be found and in the South Shetland and South Orkney Islands that are geographically part of Antarctica itself. Whaling required more specialised equipment and skills than sealing, and had a higher social status in the home ports. Whalers, especially in the mid-nineteenth century, sometimes hunted elephant seals to augment their whale oil catch, elephant oil being of high quality and readily available on some islands. When whaling was modernised in the late nineteenth century it became a highly capitalised industrial process, quite unlike the small-scale open-boat whaling that preceded it. The earlier form of sailing ship and open whaleboat whaling was common in the sub-Antarctic waters, but not so in the South Shetlands.

The rendering of penguin carcasses for their oil appears to have been a largely domestic-scale opportunistic activity on a number of islands such as Marion Island and Tristan da Cunha in the nineteenth century (Hince 2000: 258–259), but in the late-nineteenth and early twentieth centuries, it became an industry limited to the Malvinas/Falkland Islands, Macquarie Island, and possibly on South Georgia, though no physical and little historical evidence of it has been reported there. It utilised steam digesters that were initially introduced to render elephant seal oil, as an opportunistic industry alongside elephant seal oiling, and became more dominant after elephant seals became locally scarce (see Cumpston 1968; Headland 1984: 47, 53).

Sealing on the sub-Antarctic islands and the continental coasts bordering the southern oceans has a long and inter-related history (see Table 1.1 for more information). Our knowledge of the human interaction with and utilisation of most of these islands comes partly from historical sources and partly from archaeological surveys and studies aimed at gathering information for sites management and conservation planning. The extent of research archaeology – asking questions about human activities broader than those needed for planning and conservation – has been limited, but most apparent in the South Shetland Islands research (see Chapters 3 and 4). The role of archaeology in the study of sealing, whaling and penguining sites in various locations is outlined below.

2.4.1 Macquarie Island

Macquarie Island was discovered in 1810 by British captain Frederick Hasselborough, the master of the sealing brig *Perseverance* based in Sydney. A sealing gang was immediately landed while the *Perseverance* returned to Sydney to inform the owners and secure an additional sealing gang. Initially sealing concentrated on fur seals, but these became scarce within five years and extinct within 15 on Macquarie Island, and elephant sealing became the mainstay. Sealing became sporadic, however, with only three ships visiting between 1830 and 1873, before a revival in elephant sealing occurred. In 1887 Joseph Hatch, a New Zealand entrepreneur, set up operations at several points on the island, introducing large steam digesters, first for elephant oil, then for penguin oil as well (Cumpston 1968; Townrow 1989).

Archaeological investigations of a number of these sealing locations were undertaken in the 1980s. Australian archaeologist Martin Davies carried out a preliminary survey of sealing and penguining sites on Macquarie Island in 1982. In summarising the research potential of the work, Davies (1982) wrote that:

> The archaeological record contains information relating to past activities of humans on the Island. From a study of these cultural remains one can examine human behavioural patterns and when seen in the context of early European settlement on other islands and on the Australian mainland one can assess the adaptations which were necessary to exist on Macquarie with its unique climatic conditions and isolation. Also by virtue of these conditions there is the high probability that organic remains are well preserved, a situation rarely found elsewhere in Australia.

Davies thought that the study of elephant seal biology and behaviour on Macquarie could help interpret prehistoric sites in southern Australia where elephant seals previously existed but are now extinct. He also felt that the study of the architecture of huts utilising local materials, with limited imported materials, could focus on the spatial and temporal distribution of structures, illuminating their use or origins, and that the study of sites and the documentary record in combination might show differences between sites that would require re-interpretation of the historical process (Davies 1982).

Davies' initial work was followed up in 1986–88 by archaeologist Karen Townrow, including a full year of work on the island (Townrow 1989, 1990). The project was motivated by the need to inventory the sites to ensure effective management of the island (a Tasmanian Government Nature Reserve and subsequently a World Heritage Site), and involved a very thorough survey of sealing and penguining sites including ten excavations of key sites, and the study of the Australian scientific research base established in 1911 at the Isthmus and continuously occupied since 1948 (Townrow 1989) (Figure 2.21). The objective of the excavations was '… to provide optimum information on the lifestyles of the early inhabitants which is not available in historical documents' and to excavate a range of sites which 'constitute the nucleus of activity, such as living quarters or production sites' (Townrow 1989: 1).

While sealing on Macquarie Island commenced in 1810, the difficulties in locating early sites in a very dynamic environment, with subsequent disturbance by later sealing and penguining parties, by elephant seal wallowing, and by landslides and soil erosion, meant that few artefacts or sites could be definitely associated with the pre-1870s period after which occupation expanded

FIGURE 2.21 Penguin digester factory at Nuggets Point, Macquarie Island. Archaeologist Karen Townrow to right with timber from the collapsed digester building on the ground amidst elephant seal wallows (Photo by Michael Pearson 1987).

(Figures 2.22 and 2.23). The excavated archaeological evidence casts light on the lifestyles of sealers and penguin plant crews, and particularly on the type of materials imported for their sustenance. Townrow observed that the artefacts excavated from the sites showed a broader range of goods and equipment than implied by the ships' inventories and stores lists (Townrow 1989: 22–27). The gradual increase in the use of imported sawn timber as opposed to timbers from shipwrecks in penguining structures is interpreted as reflecting the increasing investment in longer-term operations between 1902 and 1920 by the leasee Joseph Hatch and a resulting improvement in living conditions for workers (Townrow 1990: 29).

Townrow also took samples from timber found on the beaches and from early hut foundations, to test the proposition that the timbers might have originated from the eight known shipwrecks on the island. While not able to link timbers to specific wrecks, species identification allowed general conclusions about which local wrecks might be the source. Of particular interest was the finding that all of the numerous barrel staves and head pieces found were of non-Australian species, suggesting the possibility that Australian timbers were not suitable, or were not utilised, for barrel production, a proposition of interest to Australian history and archaeology that has yet to be fully researched (Townrow and Nash 1990). Archaeological remains of a fur seal skull were also investigated to try to determine the species of fur seal that originally occupied Macquarie Island before

FIGURE 2.22 Penguin digester at the Isthmus, Macquarie Island, showing the damage done to the timber floor of the digester building by elephant seal wallowing (Photo by Michael Pearson 1987).

FIGURE 2.23 The same site as 2.22 four years later, after the construction of heavy fences to keep seals off the site – ugly and only effective in the short term and later removed (Photo by Michael Pearson 1991).

being made extinct about 1820 by sealing, but it proved to be a vagrant Antarctic fur seal (*Arctocephalus gazella*), not the more likely original species (*A. tropicalis, A. pusillus* or *A. forsteri*) (Townrow and Shaughnessy 1991).

The continuation of sealing and penguining on Macquarie Island was actively campaigned against by Sir Douglas Mawson (who had maintained a party on the island as a radio relay station during the AAE 1911–14), Sir Baldwin Spencer, photographer Frank Hurley and others following the First World War, with support for the creation of a nature reserve coming also from prominent members of Britain's 'heroic era' expeditions (Cumpston 1968: 307–316). Taken up by Australian politicians at national and state levels, the campaign resulted in the withholding of the license held by Joseph Hatch to operate the oiling operation, and the liquidation of Hatch's company in 1920. The island became a wildlife sanctuary in 1933, a Tasmanian State Reserve in 1978, and was inscribed on the UNESCO World Heritage List in 1997, its protection being guided through a management plan (Nash 2000; Parks and Wildlife Service 2006).

2.4.2 Heard Island

Heard Island was officially discovered by American Captain John Heard in the ship *Oriental* in November 1853, and sealing began with the arrival of Captain Erasmus Darwin in the *Corinthian* in March 1855. It has been hypothesised that the island may have been discovered and sealed before that date with no record surviving (Downes and Downes 2006). Some 40 vessels made 104 voyages seeking elephant seals on Heard Island in the 30 years after 1855. Of 79 voyages published by Downes (1996), other than one from Hobart and two from Cape Town, all originated in New England, United States, 62 of them from New London. An External Territory of Australia, Heard Island became a UNESCO World Heritage Site in 1997, and is managed in accordance to a Marine Reserve Management Plan, all actions potentially impacting on heritage sites being subject to an impact assessment and permit system (Commonwealth of Australia 2014).

An archaeological survey of Heard Island was carried out in 1986–87 by Australian archaeologists Estelle Lazer and Angela McGowan (1990), as part of planning the long-term management of the island and prioritising actions for threatened sites. The survey was also to establish a basis for an ongoing archaeological research program, a program that did not eventuate. Lazer and McGowan (1990: 15) hoped that

> The information gained from such an examination would supplement the written records of human activity in the Sothern Ocean as the archaeological record often provides different information to that of written sources. For example, collections of refuse give insights into the daily lifestyles of people that are never considered in contemporary documents. Further, archaeological data can be used to test the accuracy of historical material, an especially useful exercise considering the tendency to secrecy amongst 19th century sealers who were often concerned about protecting the exclusivity of sealing sites.

The Heard Island surveys and excavations have recorded 22 try-pots (three labelled 'Coalbrookdale'), a copper cooling tank, a scrap blubber press (a similar one having been found on Kerguelen Island) and component parts of another, a copper blubber skimmer, blubber fork, hooks, a lance, flencing tools, grindstone and many barrels associated with try-work sites (Burton and Williams 1985; McGowan 2000; Lazer and McGowan 1990). While try-pots have been reported on a number of sub-Antarctic islands, on Heard Island, the large number (22) is unusual, and some of these also survive on or are associated with try-work fireplace structures – in one of which, at North Spit Bay, three try-pots were still in their original locations abutting each other on the fireplace which was designed for five pots (Burton and Williams 1985: 51). Other try-pots have been reported on Macquarie Island (Townrow 1989), South Georgia (Headland 1984), Kerguelen and Crozet (Le Mouël 2004, Laetitia Therond (TAAF) pers. comm. 2020) and Prince Edward Islands (Cooper and Avery 1986; Graham 1989).

Thirty two sites were recorded on Heard Island, and selected condition reports were made of sites informed by the off-site guidance of conservators Janet Hughes and Linda Clark (Lazer and McGowan 1990: 20). Salvage excavation was carried out at Oil Barrel Point due to the retreating coastline threatening a double line of 35 oil barrels still containing traces of elephant oil (Figure 2.24), the most threatened barrels being excavated and removed for conservation and display. A study of lichens was undertaken as a potential dating technique using their growth over artefacts as an indicator of age. It was concluded, however, that there were many variables effecting lichen growth in particular locations, and that a long-term experimental program would be needed to calculate and account for the various factors that might accelerate or constrain lichen growth rates (Lazer and McGowan 1990: 21).

Based on the observation made during the survey, McGowan expanded on her analysis of the sealers' remains (McGowan 2000). Stone platforms appear to be a form of sealing architecture unique to Heard Island, not being identified at any other reported sealing location. Five such mounds were located, constructed of large beach cobbles in a semi-circular mound between 4 and 17 m across and about half a metre high. These were located on plains slightly inland of the shore in areas heavily dissected by seal wallows, the stone platforms providing elevated areas above the disturbed ground: 'The function of the platforms was apparently as working platforms and the more intact ones supported remains of try-works including try-pots, flensing floors and other equipment' (McGowan 2000: 63).

Residential sites were also found, with evidence of low stone walls, the use of tussock turves and planks as wall materials, and upright wooden poles to support canvas roofs (Figure 2.25). Some were located in the lee of large boulders.

FIGURE 2.24 Oil barrels eroding from a double line of 35 barrels, abandoned at the site in 1882. Coastal erosion, exacerbated by climate change, is a major threat to sealing sites (Photo by Kevin Kiernan 2000).

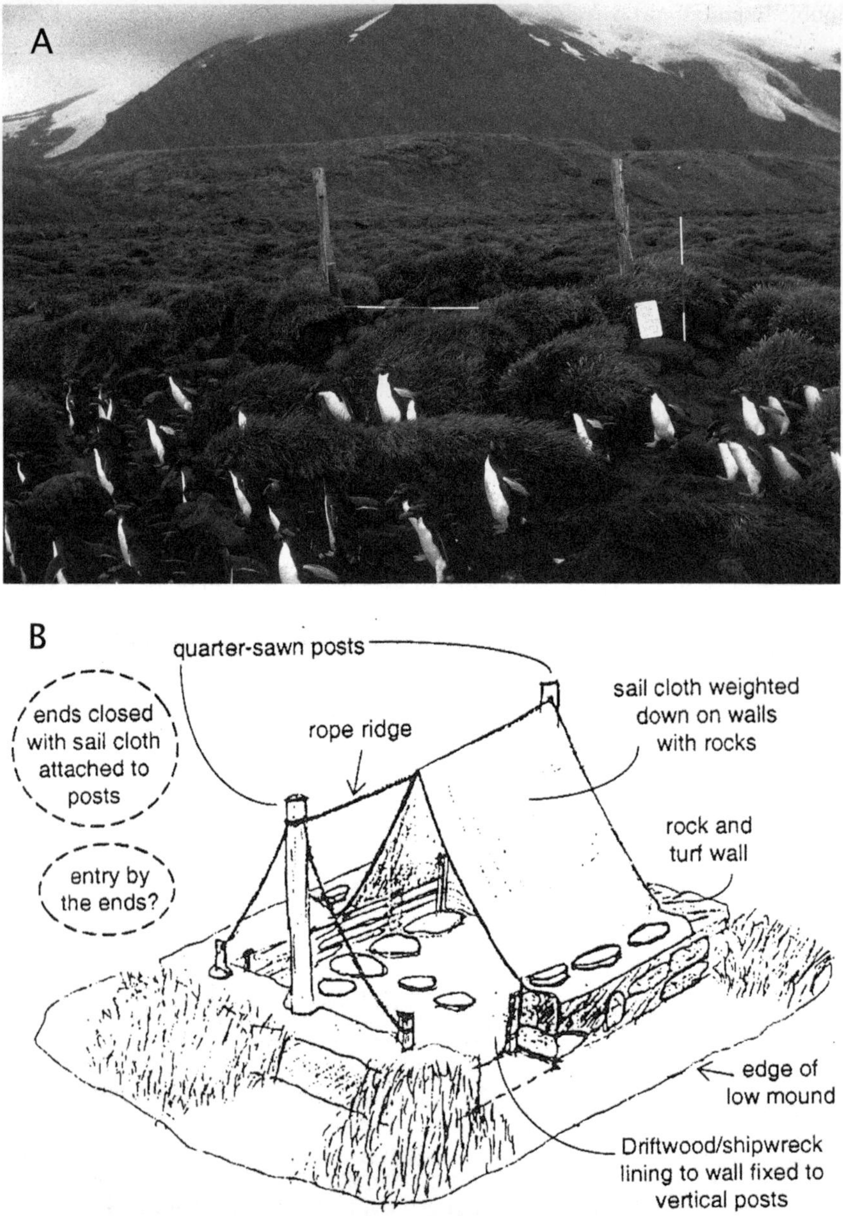

FIGURE 2.25 Long Beach, Heard Island: (a) Two standing posts with a low rock and turf wall, for a tent-hut (Photo by Angela McGowan 1986); (b) Reconstruction of the sealers hut following archaeological information (Source: McGowan 2000: 65).

McGowan (2000: 65–67) comments that from the archaeological evidence the huts appear to be better constructed and more liveable than suggested by the documentary reports, which stress miserable makeshift shacks. Lava caves had also been occupied. Early reports also recorded pits in the ground covered with wooden boards used as living huts, and archaeologist Anne Robb found such a pit hut (Robb 1988: 155). Some patterns seemed to distinguish different groups or periods of sealing. In the northwest of the island, the work sites and domestic sites were close together and sometimes indistinguishable one from the other. The variety of different building styles observed might suggest separate ventures, perhaps coming from different ships, in different years, or competing gangs. On the other hand, at the Spit End at the southeast end of the island, there seems to have been a greater separation between work and living areas, the former being on platforms near the coast (described above), and the latter on the moraine further inland and above the worst of the seal wallows. The similarities among these sites suggest that they almost certainly originated from the same or related operations. They have clear and consistent separation between work and living areas; the try-works show a strongly shared style and all were on stone platforms. It may be that they represent the last operations on these beaches, overlaying as they do earlier sites. The last documented expedition to this area was in the 1880s when the *Trinity* was wrecked, leaving the crew marooned for 15 months (Oct 1880-Jan 1882 – see Downes 2010). The *Trinity* party worked over two seasons, with diminishing cooperation from the crew in appalling conditions, gathering oil until finally rescued. The Oil Barrel Point barrels (mentioned above), which contained the gathered oil, were left abandoned at the end of that enforced stay because the US naval ship that rescued the sealers would not take them aboard (McGowan 2000: 68; Downes 2010; McGowan and Lazer 2017).

McGowan (2000) observes that:

- The artefact assemblage could provide a basis for the study of technological change in sealing over an 80-year period in the sub-Antarctic.
- The documented social and racial tensions and the separation of American and Cape Verde Islander (African) populations of sealers may be reflected in the spatial arrangement and use of sites, identified by ethnically determined artefacts. Comparison with other racially diverse sealing populations might then be possible. Cape Verde Islanders made up about half of the crews of American sealers at Kerguelen and Heard Islands after the American Civil War (Busch 1980a: 124).
- The sites are potentially useful in studying the broad questions of the organisation of the sealing industry and associated networks of technology and commerce on a global basis.

McGowan argues that sealing on Heard Island should not be interpreted as a colonisation event, due to the short-term exploitative nature of the industry and the absence of long-term aspirations of use or control of the land (McGowan 2000: 69).

The Heard Island sites are rapidly disappearing due to sea-level erosion, changes in ice cover, and associated erosion linked to climate change and seal activity. Systematic research such as that recommended in Lazer and McGowan's original report (1990) is urgent.

2.4.3 Kerguelen Island and the Crozet Archipelago

Kerguelen Island was discovered in 1772 by Yves-Joseph de Kerguelen Trémarec, and charted by Capt. James Cook in 1776. While Cook named the island after Kerguelen, his alternative name for the island, 'Desolation', was consistently used by American sealers throughout the nineteenth century. The Crozet islands were also found in 1772, by Marc-Joseph Marion-Dufresne. Sealing for fur seals commenced on Kerguelen in 1792, and on the Crozets in 1804, mainly by Americans (Busch 1980a: 119; Busch 1985: 26). Fur seal numbers dropped at both islands after a decade or so of sealing, and visits became sporadic until elephant sealing began in 1832 by South African sealers on the Crozets, and by American ships on Kerguelen by 1838–39 (Decker 1973: 131; Richards 1991; Richards 2010: 56). From the 1830s–40s, trying out of blubber increasingly took place aboard ship (Le Mouël 2004: 61). Vessels commonly were based at Kerguelen and dropped gangs on Heard Island, 450 km to the southeast, for the season, and from 1840 to 1890 one in four whaling vessels leaving New London (the major source of vessels at these islands at the time) also called at Kerguelen and Heard Islands for elephant seals (Busch 1980a: 121, 125). A modern whaling station was built at Port Jeanne d'Arc in 1908, closing in 1929. The remaining buildings at the station were recorded, and one set of standing buildings was excavated and restored in 2000 (Le Mouël 2004: 63–64) (Figure 2.26). The whaling station at Port Jeanne d'Arc was digitally recorded in 2010, and 3D models are available on the Terres australes et antarctiques françaises (TAAF) website.

French archaeologist Jean-François Le Mouël commenced an archaeological survey of the sites on Kerguelen and Crozet in 1993–94 (Le Mouël 1994). This survey located sealing and whaling sites, including try-works used for seal and whale oil rendering, with heaps of barrel hoops, chains and ship components marking land depots for ship-based whalers (at L'ile-aux-cochons, Hog Island), and a cemetery, whale skeletons and sealer hut sites (at Betsy Bay). A cave with stone walls built inside for protection from the weather and a range of artefacts were found at La Grotte des Phoquiers. At Pointe Charlotte (aka Long Point), the site of the sealer hut 'Hope Cottage' – which features in detail in John Nunn's published memoir of his shipwrecking on Kerguelen in 1825 (Clarke 1850) – was excavated in 1996, but little artefactual evidence remained (Le Mouël 1996: 30–77). A small cave site was also located, with stone walls built in front for protection from the weather. This is interpreted as a shelter used by sealers shipwrecked from the ship *Hebe* on the island in 1831, named 'Grotte-refuge de la Hebe' by Courbon (2012). The site has not been excavated (Le Mouël 1996: 78–82).

On the Crozet Archipelago, on Possession Island, a major elephant sealing try-works was identified at American Bay, in what is now named Vallée des Phoquiers (The Valley of the Sealers) (Le Mouël 2006; Courbon 2012; Le Mouël et al. 2013; Motte and Vidal 2020). This consisted of a single try-pot embedded in a furnace and fireplace built up of local basalt boulders, with the clearly defined foundations of three hut sites located in the valley immediately above the try-works. Unfortunately, the try-works was washed by the sea at times, and it was judged necessary to record and remove the structure to higher ground in 2006, protecting it in the meantime with sandbags (Figure 2.27). During excavation and removal of the try-works, residues of the blubber fuel used were found, and a chimney housing was located behind the try-pot. All stones were numbered, photographed and surveyed for re-erection. The try-pot was removed to France for conservation treatment, with the intention of returning the pot to the island for re-erection, though this has not yet occurred at the time of writing (2021). The 3D digital representation of a number of try-pots and of the Vallée

FIGURE 2.26 Port Jeanne d'Arc whaling station, Kerguelen. Buildings at rear excavated by Jean-François Le Mouël and conserved (Source: Franek2, Panoramio, Creative Commons).

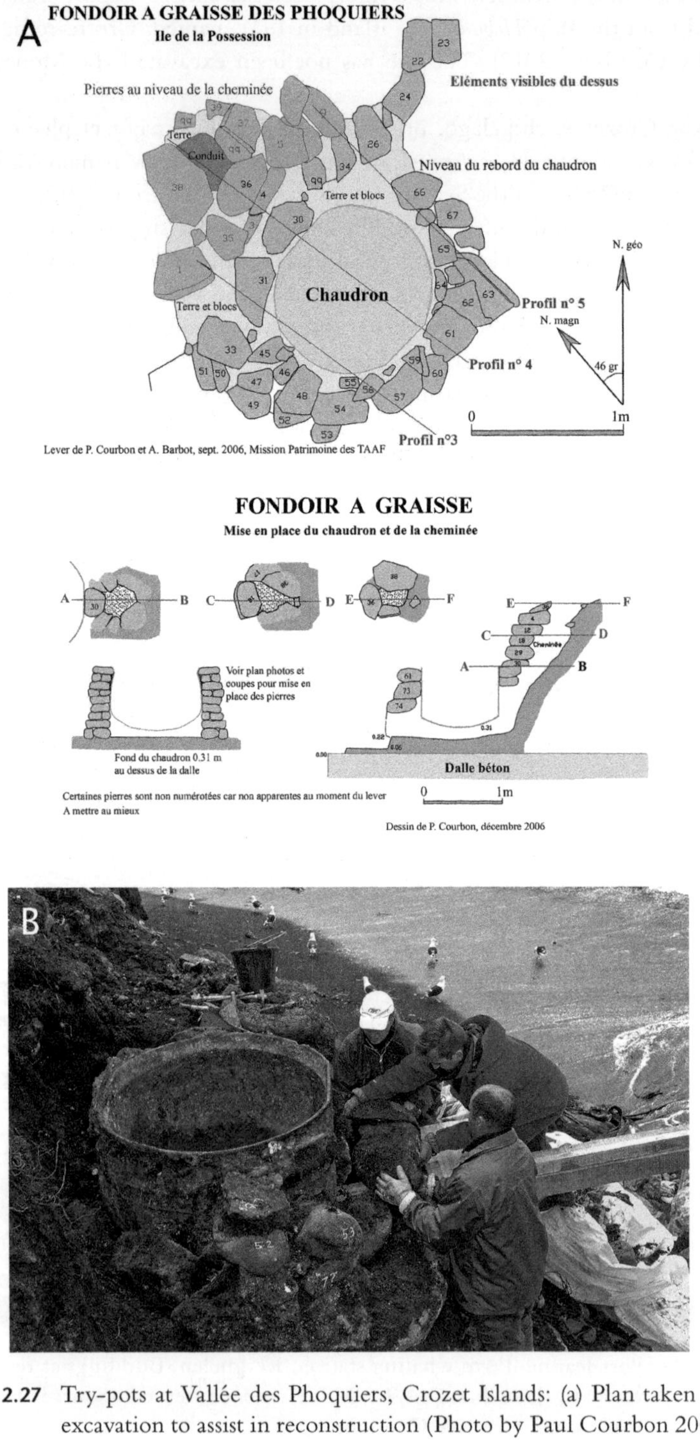

FIGURE 2.27 Try-pots at Vallée des Phoquiers, Crozet Islands: (a) Plan taken during excavation to assist in reconstruction (Photo by Paul Courbon 2006); (b) Excavation of try-pots (Photo by Jean-François Le Mouël, TAAF 2006).

des Phoquiers site has been made available on the web (see https://sketchfab.com/Archeologie_ol/collections/taaf).

2.4.4 Prince Edward Islands

Discovered in 1663 and re-discovered in 1772, the Prince Edward Islands were first landed on sometime between 1799 and 1802, probably by the sealers Pierre François Péron and Stephen Hall in the *Sally*, closely followed by Henry Fanning in the *Catherine* of New York in 1803 or early 1804 (Graham 1989: 6–11; Richards 1991 and 2010: 24; Headland 2009: 105). This was followed by a massive slaughter of fur seals in 1805–06, after which sealing visits were sporadic, though large takes of elephant seal oil are recorded after 1810, such as in the published memoirs of William Phelps recalling his two-year stay in 1818–20 (Phelps 1871). The Prince Edward Islands are dependencies of South Africa, and work there is managed by the South African National Research Foundation's South African Antarctic Programme (SANAP).

A compilation of known sites on the Prince Edward Islands was produced in 1986, listing 18 sealing or shipwreck sites on Marion Island and six sites on Prince Edward Island (Cooper and Avery 1986), and further surveys occurred in the 1990s (Boshoff et al. 1997; Boshoff and Schalkwyk 1999). Archaeological survey and excavation has taken place at a survivor camp from the 1908 shipwreck of the sealer *Solglimt* on Marion Island (Figure 2.28), first by Tom Graham in 1989, then by Tara Van Niekerk in 2013, associated with the first successful major investigation of an underwater shipwreck site in Antarctic waters carried out on *Solglimt* in 2011 and 2014 (Graham 1989; Boshoff et al. 2015; Van Niekerk 2016; Cooper et al. 2018). The wreck itself has only had a preliminary recording carried out, but potentially contains more information on the supply and outfitting of an early twentieth-century seal hunting expedition. 'The *Solglimt* was regarded as one of the best-outfitted vessels of its era although no details have been found in any manifests or other sources as to what this entails' (Boshoff et al. 2015: 58). The archaeology of the ship and of the survivor camp may help expand that knowledge.

The work on the survivor camp focused on clarifying the extent of remains from the 1908 occupation, which was spread between a beach, two cliff-face caves and a valley leading down to the sea, to try to understand how and perhaps why the camp was organised in this way (Van Niekerk 2016: 39). The documentary record, consisting of official and unofficial records of the wreck, sealers' logbooks, and photographic documentation, was studied to identify what evidence it could provide of camp arrangements and organisation. The archaeological survey was informed by methodologies used at a number of other shipwreck survivor campsites around the world (see earlier in this chapter). Three areas were excavated to varying degrees. It was appreciated that a knowledge of the cultural norms and practices of the time, which might have influenced the behaviour of the survivors, was an essential component in interpreting the archaeological evidence.

FIGURE 2.28 Photograph of the wreck of the *Solglimt* washed by heavy seas, on Marion Island, possibly taken by Captain Anders Harboe-Ree, 1908 (Photo from Cathrine Harboe-Ree Collection).

The models of salvage and disaster-response discussed by Martin Gibbs (2006a, 2006b) and Clark and de Biran (2010) were applied to the analysis of the sites (Van Niekerk 2016: 76). Four main activities were identified with both documentary and archaeological expression: subsistence (salvaging food, fuel, clothing, tarpaulins, wood and other provisions from the wreck); habitation (firstly a house to shelter the provisions, a cookhouse with stove, then shelters for themselves in huts and rock overhangs); development of an escape strategy (strengthening and decking one of the three surviving boats, in which to send a relief party to South Africa); and defence (from internal dissention) (Van Niekerk 2016: 77–82). The close proximity of the shipwreck site to good habitation sites was a key factor in the location of the camp. The separation of the work area from the habitation area and the central location of the cookhouse (identified by its large size, stoneware and cooking stove remains) is tentatively suggested as mimicking ship-based spatial relationships (Van Niekerk 2016: 84). The separation of the officers and crew on the basis of status (the officers, engineers and doctor separated from the crew) is mentioned in the documentary record, and echoed in the construction of multiple and separated huts. One aspect of the models for other survival camps that did not appear in the archaeological record was the introduction of a subsistence strategy to gather local food and resources – it would appear that sufficient ship's stores were landed to support the survivors until a rescue ship arrived just a month after the wreck (Van Niekerk 2016: 87).

The particular physical and social conditions associated with shipwreck survivor camps are repeated at shipwreck sites elsewhere in the sub-Antarctic and Antarctica (such as several 1820s shipwreck sites in the South Shetlands, known from the sparse documentary record to have had survivor camps and sites for constructing escape vessels associated with them). However, none of these have as yet been studied archaeologically.

Work on the sites in the Prince Edward Islands is continuing under the *Archaeologies of Antarctica Project* (Iziko Museums of South Africa) and, with a wider focus on heritage objectives, the *Antarctic Legacy of South Africa Project* (in the South African National Antarctic Programme).

2.4.5 South Georgia

Sealing on South Georgia commenced in 1786, with traditional non-industrial sealing ceasing in 1913, while elephant sealing was undertaken at Grytviken in conjunction with the whaling operation there from 1909 until 1964 (Headland 1984: 32–57, 126–130). The South Georgia Heritage Trust, in partnership with archaeologists from Cambridge Archaeology Unit, commenced the South Georgia Archaeological Project in 2019, surveying and test excavating a number of eighteenth- and nineteenth-century sealing sites around the island, as part of an overall program to study and conserve the island's heritage (Cordeiro 2019; Neil 2019; South Georgia Association 2021). Thirty sites connected with sealing have been identified, the project studying 15 of these in 2019. The project aims to progressively record and study the other sites to seek information on the operation of the elephant and fur seal industries and to study the living conditions of the sealers. Managed by Robert Burton with field work led by archaeologist Marcus Brittain, the project is ongoing, and while many samples have been collected for residue and chemical analysis, no research results had been published at the time of writing. A major issue is that the resurgence of seal populations to their original numbers has removed much of the tussock cover that developed after the initial sealing era and protected the archaeological sites. This development makes the survey and study of the remaining sites an urgent research priority.

Modern Antarctic whaling started in South Georgia in 1904, with six shore stations being established there, operating in different periods until 1965, after which the stations were abandoned and commenced to rapidly decline in condition (Headland 1984; Basberg et al. 1996: 51; Basberg 2004). The South Georgia whaling stations were an integral part of the growth of industrialised whaling in the southern oceans, which saw the commercial potential of marine species exploited to an unprecedented extent, and the rise of international science and politics in Antarctic waters in equal measure (Burnett 2012). An industrial archaeological study of the whaling stations on South Georgia from 1990, led by Bjørn Basberg, was aimed at accurately recording the station buildings, associated grounded ships, machinery and tools, and other elements of the stations before they deteriorated further. 'A subsidiary aim was to undertake the survey in a manner that benefitted its potential numerous cross-disciplinary perspectives' (Basberg et al. 1996: 51). Two of the large and complex stations, Husvik and Stromness, were surveyed by Dag Nævestad and Basberg in 1990; that at Grytviken was surveyed in 1993 by Basberg and Gustav Rossnes; and Leith Harbour in 1996–97 by Basberg, Nævestad, Rossnes and Gisle Løkken (Nævestad et al. 1996). The whaling station at Prince Olaf Harbour was visited briefly in 1990 and 1999 but not surveyed in detail, and that at Ocean Harbour was analysed from earlier maps and material (Basberg 2004: 15–21, 68). A laser scan survey of Grytviken, Husvik and Stromness was undertaken in 2012 by Russell Gibb and Daniel McCurdy to supplement the Basberg survey with digital recording, followed up by three more expeditions that recorded all the whaling stations (Gibb and McCurdy 2013; Russell Gibb pers. comm. 26/06/20).

Louwrens Hacquebord and Dag Avango, in an overview of industrial sites in polar regions and the Large Scale Historic Exploitation of Polar Areas (LASHIPA) project, have concluded that sites such as the whaling stations on South Georgia, in the South Shetland Islands and whaling harbours on the Antarctic Peninsula, can provide unique evidence. This includes factors driving the industry, the design of whaling technology, the structure of settlements and the strategies to ensure the catch. In these cases, the evidence available through survey and excavation of sites of the whaling industry gives additional and often different information to that available from written sources, enabling a more holistic and comprehensive understanding of the whaling activities (Hacquebord and Avango 2016).

Each station combined a large industrial whale reduction and processing plant, and the equivalent of a small town with worker accommodation, messes, hospital, bathhouse and church (at Grytviken only). Domestic life was supported by provision stores, bakery and butchery; pens for domestic animals such as cattle, sheep, pigs and chickens; and slop chest where clothing, tobacco, sweets, etc. were available. The business of whaling was supported by workshops and repair facilities for the plant and the associated ships, all powered by hydro-electric, diesel and steam power generation plants. Recreational facilities included library, cinema, soccer field and ski jump. The extent of servicing and recreational infrastructure is a physical expression of the whaling companies' efforts to keep workers satisfied, and so reduce the likelihood of the crippling impacts of industrial strike actions at stations where there was no alternative labour source (Hacquebord and Avango 2016). It was an adjustment of the methods of industrial capitalism crafted to suit the particular social isolation of the sub-Antarctic, the sail-in-sail-out nature of the labour supply, and the scale and complexity of the modern whaling enterprise.

Given the vast extent of these establishments, Basberg's team prioritised documenting the layout and the pattern of functions, the use and functions of each building and room, and reconstructing the stages in the industrial process so they could be more easily understood. The history of the technology of modern whaling had already been well documented in surviving construction drawings and patents, but the more human aspects such as '... the stations as functional backgrounds and conditioners of social life...' were less well understood (Basberg et al. 1996: 52). In part, this aspect was addressed by subsequently interviewing former workers of the stations, armed with the detailed recordings of each station with which to stimulate memories and track personal movements and activities. It was recognised that the recordings represented the stations at the end of their working lives, and as such provided an as-completed baseline understanding of each complex. This has the potential for use in researching the evolution of the stations over time based on both material evidence and documentary evidence such as plans, maps and photographs.

A preliminary study of the living conditions and social system of Husvik Harbour was undertaken, using the survey material and subsequent oral history interviews, but it was acknowledged that this showed the promise of such work rather than fulfilling its full potential. One of the conclusions was that the social structure was more like that of a ship than that of an industrial plant. The accommodation and messing arrangements indicated a highly stratified society, again reflecting the situation aboard ship. A divergence from this model was finding a number of sleeping areas created by men with sufficient status to occupy lofts or attics of buildings used for other purposes and to furnish them to their own taste (Basberg et al. 1996: 59–62; Basberg 2004: 158–159). The creation of this hierarchy has been interpreted as demonstrating the great importance placed by the whaling companies on maintaining social control over a large and isolated workforce (Hacquebord and Avango 2016). The industrial organisation of the stations was also informed by the surveys, allowing the size of particular processing areas (such as blubber plant, bone plant, meat plant and guano plant) to be calculated and compared station to station, and station to pelagic whaleship arrangements (Basberg et al. 1996: 62–64).

The studies of the South Georgia whaling stations provided an excellent example of the relative strengths and necessary combination of textual and material evidence together with intangible evidence in the form of oral history. It also demonstrated the reality that the actions of individuals (such as the creation of individualised living spaces) can be discerned in even the most structured and organised industrial setting. While they undoubtedly have research value, the surveys might also represent the last opportunity to record and understand the whaling factories in a more-or-less intact state, as their condition continues to decline and as clean-up activities likely become recurrent necessities in coming decades.

2.4.6 New Zealand Sub-Antarctic Islands

2.4.6.1 Auckland Islands

First occupied or visited by Polynesian voyagers between the thirteenth and fifteenth centuries AD, as part of the exploration south of New Zealand also reflected in Māori tradition (see Wehi et al. 2021), the Auckland Islands were re-discovered in 1806, and sealing commenced very soon after by sealers working mainly out of Sydney and Hobart. Seal stocks dwindled by 1812, but episodic sealing followed in the 1820s. One of the sealing ships in 1823, the schooner *Henry* of New York, Captain Robert Johnson, is very likely to have been the same ship that was sealing in the South Shetlands for two seasons 1820–22 – where a Robert Johnson (one of several named in the 1820s fleets) was captain of the *Jane Maria* which sealed in the same fleet as the *Henry* (then under Captain Brunow) in the same years (Headland 2009: 130, 139; Prickett 2009: 43). Archaeological evidence of sealing on the islands is slight, a survey in 2003 by New Zealand archaeologist Nigel Prickett identifying 12 sites as having probable sealing associations. Because the islands have a long history of European use, it is often difficult to verify that sites resulted from sealing rather than some other activity, and in the absence of excavations, identification is usually based on documentary sources. Above-ground remains at these sites commonly consist of cut tree stumps and terracing for huts, with no diagnostic evidence pointing to sealing (Prickett 2009).

A research strategy was adopted for work in the New Zealand sub-Antarctic islands in 2005 (West 2005). The strategy called for:

- survey and mapping of pre-European Polynesian sites on Enderby Island, backed up with core sampling and test pitting to establish the extent of the archaeological remains (some of which is reported in Dingwall et al. 2009);
- detailed archaeological investigation of some of the recorded sealing sites in the Auckland Islands, to enhance knowledge about this aspect of the Island's history which is not well recorded in documentary sources;
- archaeological survey of sites already actively managed, such as Camp Cove and the North East Harbour whaling station on Campbell Island; and
- further research and survey on Campbell Island, Antipodes Islands and Snares Island, where current information was basic and largely based on literary references

It was pointed out that the 2003 survey on the Auckland Islands had increased the number of recorded historic sites of all types from 53 to nearly 130. The strategy suggested that recording work was also needed for the remains of shipwrecks, and that the nature of cultural landscapes on the islands had to be investigated. Implementation of the strategy appears to have been limited as yet (though see Petchey et al. 2015).

Also surveyed in the 2003 expedition was the site of the 1849–52 Enderby settlement associated with the British Southern Whale Fishery at Erebus Cove, Port Ross, on Auckland Island (Dingwall and Jones 2009). The 30-odd timber and iron buildings were taken apart and removed when the settlement was abandoned, with only one house left standing. However, sufficient surface evidence was identified to be able to reconstruct the layout and features of the settlement.

While not associated with sealing or whaling, there have been archaeological surveys of the land-based evidence related to some of the ten known shipwrecks on the island. The crews of seven of these wrecks survived in shore camps for from three to 20 months before being rescued. The five-man crew of the 1864 wreck of the schooner *Grafton*, lost during a storm at Carnley Harbour, built a boat in which three of them sailed to Stewart Island and arranged for a ship to be sent to rescue the two remaining men after 20 months ashore. The archaeological survey in 2002 identified a boat run on the beach where an attempt was made to haul the wreck ashore, and probably where the rescue boat was built; timbers from the *Grafton* wreck, together with the site of a large hut ('Epigwaitt') built from wreckage, pathways, a forge, evidence of forest clearing and other features (Egerton et al. 2009). Less extensive evidence of a number of other shipwreck campsites was identified, together with a range of government-built castaway shelters, message posts and boatsheds.

2.4.7 Whaling sites in the South Shetlands and Antarctic Peninsula

In the area that is now managed as 'Antarctica' under the Antarctic Treaty, south of 60°S latitude, modern whaling took place in the South Shetlands, South Orkneys and the Antarctic Peninsula. In 1905, Christen Christensen sent the ship *Admiralen* to Admiralty Bay, on King George Island in the South Shetlands, to commence whaling there. In 1906–07, Adolf A. Andresen took the whaling ship *Gobernador Bories* to the South Shetlands for the Chilean company Sociedad Ballenera de Magallanes and used Whalers Bay on Deception Island as its base. By 1909, nine factory ships were moored within Whalers Bay (then known as Factory Bay). Factory ships also operated at the South Orkneys from 1912. Some ventured onto the Antarctic Peninsula, to Port Lockroy on Wiencke Island, Neko Harbour on the Danco Coast and other locations. By World War I, 12 factory ships were operating in the South Shetland region (Basberg 2007). The wreck of the whale factory ship *Guvernøren* (or *Gouvernøren*), lost due to an onboard fire while at her mooring in 1915, rests partly above water at Foyn Harbour, Enterprise Island, on the Antarctic Peninsula (Headland 2009: 265). The crew of 85 was rescued by another whaleship. A whale chaser, *Southern Hunter*, was wrecked at Neptune's Bellows entering Deception Island's caldera in 1959 (Headland 2009: 350), and sections of the hull lie on the shore at the entrance (Figure 2.29). In 1924 the chaser *Bransfield* sank in South Bay, Livingston Island, and another, the *Graham*, sank near Bridgeman Island with the loss of ten men (Headland 2009: 277). No maritime archaeological work is known to have been carried out at any of these wreck sites.

At Admiralty Bay, on King George Island in the South Shetlands, whaling was always ship-based, the bay being used for shelter while processing whales, and for watering the ships. A survey of the remains of whale bones and whaling equipment along the shores of Admiralty Bay was conducted in 1996–98, but it was not possible to correlate the observed remains with specific whale

FIGURE 2.29 Wreck of the Southern Hunter (1959) on rocks at Neptune's Bellows, the narrow entrance to Foster Harbour at Deception Island (Photo by Michael Pearson 2008).

anchorages and operations, suggesting that for ship-based whaling documentary sources are going to provide the major source of information (Kittel 2001).

A shore-based whaling station was established at Deception Island in 1911 by the Hvalfangerselskabet Hektor AlS of Tønsberg in Norway (registered under the name Hector Whaling Ltd in London) and known as the Hektor (or Hector) Station (Figure 2.30). Another shore station was set up on Signy Island in the South Orkneys in 1921. Approximately 150 people worked at Hektor Station each summer, until its abandonment in 1931. The British reoccupied the station as 'Base B' during Operation Tabarin in 1944, and inhabited it until 1969 when the scientific station was partially destroyed by a volcanic eruption. It is now one of the most visited tourist sites in Antarctica. While Hektor whaling station was not as large as those on South Georgia, the station still had a carpenter's shop, a blacksmith's shop, a cooper's shop, an engineer's shop, a toolshop, a grindery, a laboratory, a plating shop, a powder magazine, two storehouses, four whale oil tanks, three fuel oil tanks, three water tanks, a jetty and a floating dock, as well as accommodation for the crew in a barrack, a mess and kitchen, the magistrate's house, a piggery, a radio-shed and a hospital. A graveyard with 35 graves was destroyed and buried under volcanic sediment by a volcanic mudflow in 1970 (Headland 1984; Hacquebord 1992). Louwrens Hacquebord carried out a survey of the site and an environmental study of human impacts in 1991–92 (Hacquebord 1992). The remains of the whaling station and scientific occupation period are now in ruin, and are not going to be conserved largely due to the

FIGURE 2.30 Whaling station at Whalers Bay, Deception Island, showing the living buildings, whale oil digesters, oil storage tanks and boats for carrying water to ships. The steaming beach is due to volcanic warm ground (Photo by Michael Pearson 2008).

volcanic threat. However, the opportunity has not been taken to date to undertake archaeological recording and research at the site.

2.5 Discussion of sealing, penguining and whaling archaeology

Archaeologists and others working in the Antarctic and sub-Antarctic islands recognised the potential value of archaeological evidence at sealing, whaling and penguining sites in expanding our understanding of the human interactions with the Antarctic environment. These often echoed the research values identified for Antarctic huts and scientific stations (see Section 2.3) – they were important:

- as material sources of information that either corroborated, expanded or contradicted written sources;
- as primary source information for understanding sealing and industrial whaling operations and related life experiences, especially in the context of the sealing industry, which was both poorly documented and prone to secrecy;
- as important material evidence to be understood and recorded before development, including conservation, or decay impacted them;
- as evidence to be located and understood as part of the resource to be managed by the relevant land managers.

Apart from a few examples, such as the *Solglimt* wreck survivor campsite on Marion Island and the survey work on Macquarie and Heard Islands aimed at establishing a baseline for future research, archaeology for non-conservation and management research purposes has been rare. However, much evidence has been found that addresses the research potentials so frequently postulated in the reports. Archaeological evidence has been able to show aspects of the living conditions of sealers not available in the written sources, such as the sealer shelters on Heard Island being better constructed than the documentary reports suggest, and those at Macquarie improving as the industry became more established. Patterns of human settlement and distribution such as on Heard Island and the *Solglimt* campsite on Marion Island are visible in the material remains, but not addressed at all in the written sources. On South Georgia the physical documentation of the whaling stations gives a much fuller and more integrated view of station operation and whaler life than is available from documentary sources alone. It elaborates a range of important characteristics of the whaling stations, such as the functional distribution patterning, the internal building uses and the identification of 'non-standard' practices, linking them to the oral history describing the work at the stations.

The research outcomes at shipwreck survivor campsites, like that of the *Solglimt* expands the potential for joint maritime/terrestrial archaeology associated with wreck sites and camps, utilising developments in survivor camp interpretation gathered world-wide (as raised in Section 2.3).

3

THE ARCHAEOLOGY OF THE SOUTH SHETLAND ISLANDS

As presented in Chapter 2, a significant part of the archaeological work conducted in Antarctica has focused on the conservation and study of sites associated with the 'heroic age' of exploration in the late nineteenth and early twentieth centuries. For the last decades, archaeologists have been working on the South Shetland Islands with the goal of researching the early history of Antarctica, featuring a group – the sealers – that has been relatively invisible or marginalised in the dominant narratives of the southern continent. Approaches to sealers' presence on the sub-Antarctic islands were discussed in the previous chapter. Some aspects of the exploitation and sealers' lives both in that area and on the South Shetland Islands share some features, albeit presenting peculiarities responding to each context.

Because of their close proximity to the maritime routes that connect the Atlantic to the Pacific, the South Shetland Islands – lying about 1000 km south of the tip of South America and 120 km off the Antarctic Peninsula – were the first place where humans encountered Antarctica in 1819. Furthermore, because fur and elephant seals were found there in great numbers, the archipelago was the first region of Antarctica exploited for its natural resources. Many of the shelters built by sealers during hunting operations in the nineteenth century still survive on the South Shetlands. These places make up the largest group of archaeological sites so far identified in Antarctica. The extent of the archaeological work conducted on the archipelago, as well as the emphasis placed on research questions, justifies the importance given to the region in the book.

In this chapter, we will not only discuss the history of archaeological works on the South Shetland Islands, mostly focused on sealers' presence, but we will also present in detail some of the information produced by the international project 'Landscapes in White' – coordinated by the Laboratory of Antarctic Studies in

DOI: 10.4324/9780429201257-4

Human Sciences from the Federal University of Minas Gerais (LEACH-UFMG, Brazil) – of which the authors of this volume are part. In that respect, special attention will be paid to some of the sealing shelters studied by the project and the interpretation of their uses, considering the features of the buildings and the presence of some other findings. Interpretations on the social practices connected with specific artefacts and remains will be explored in depth in Chapter 4. It is worth noting that the 'Landscapes in White' program, and the related Argentine and Chilean projects that preceded it, represent the most focused, intense and long-running archaeological research projects conducted in Antarctica to date.

3.1 History of archaeological work in the South Shetland Islands

3.1.1 Historical background to the human occupation of the archipelago

The historical background to the discovery and exploitation of the South Shetland Islands has been presented in Chapter 1 and dealt with in a number of publications (Gould 1941; Stackpole 1955; Bertrand 1971; Jones 1985; Berguño 1993a, 1993b; Campbell 2000; Pearson and Stehberg 2006; Pearson 2020). As previously mentioned, the first validated discovery of the South Shetland Islands, attributed to the British Captain William Smith in 1819, opened the way for exploitation of fur seal skins and brought Antarctica into the global fur sealing trade (see Busch 1985; Zarankin and Senatore 2005). News of Smith's discovery was soon the subject of speculation in the ports of Valparaíso and Buenos Aires and quickly reached New England and British ports. Even before Smith's discovery was officially confirmed, sealing captains based in Buenos Aires were in search of the new islands. Two ships from Buenos Aries – *San Juan Nepomuceno*, and *Espíritu Santo*, representing British merchant interests – and one from the USA – *Hersilia* – are thought to have carried out sealing in the 1819–20 summer. The surviving reports of two of the ships indicate that they took 23,468 seal skins between them (Pearson and Stehberg 2006: 336).

Calculating the number of ships visiting the South Shetlands in subsequent years is difficult, as different figures are used by various original participants and by later authors. The ship numbers presented here are collated from a number of sources (including Fildes 1820–21, 1821a; Stackpole 1953, 1955; Bertrand 1971; Jones 1985; Roberts 1958a, 1958b; Hattersley-Smith 1991; Headland 2009). The 1820–21 season was the most productive in the life of the South Shetlands as a fur sealing ground, with at least 69 vessels operating in the archipelago, the great majority coming from either New England ports or Britain. In the 1821–22 season, at least 48 vessels operated in the South Shetlands, but the damage to the seal population had already been done, with breeding colonies severely depleted and scattered. Two of the British ships came from Australia (Pearson 2018b) while several others were based in Valparaíso, Chile, and Buenos Aires, Argentina. A fleet of six ships with 85 men, sent by the New England owners Fanning and

Pendleton, was so unsuccessful that it went off to the South American islands in search of hair seals to complete its cargo (Busch 1985: 25).

Others such as British sealers George Powell and James Weddell (Figure 3.1), and American Nathaniel Palmer, went exploring for new sealing grounds (Pearson 2020). In the 1822–23 season, the number of recorded ships in the South Shetlands dropped to 12, and in the four subsequent seasons, it didn't exceed five ships (Pearson and Stehberg 2006: Table 1, 337). The attractiveness of the South Shetlands as a sealing ground was past. When natural scientist James Eights visited the archipelago in 1829, he observed of the fur seal that: 'This beautiful little animal was once most numerous here, but was almost exterminated by the sealers, at the time these islands were first discovered' (Eights, 1838: 209). There is no official figure for the total number of seal skins taken during the height of the sealing boom at the South Shetlands, but a conservative estimate of the catch for 1819–27 is 300,000 skins (Pearson and Stehberg 2006: 337).

Many of the ships sealing in the 1820s made repeated voyages, so of the known visits, 53 British ships accounted for 87 voyages, while 37 American vessels made between them 51 voyages. Only two of the American and five of the British vessels are recorded as being wrecked in the South Shetlands in this period (Figure 3.2), though a greater number of smaller tenders, called 'shallops' appear to have been lost, but records are scant (Pearson 2018d).

FIGURE 3.1 James Weddell's brig Jane (160 tons) and shallop, the cutter Beaufoy (65 tons) in the Weddell Sea February 1823 (Drawing by A. Masson, from a sketch by Captain Weddell, in Weddell 1827: facing 34).

FIGURE 3.2 Ship's figurehead, possibly from Clothier or Robert wrecked in storms at anchor in 1821 and 22, retrieved from Clothier Harbour in 1980. Institute of Patagonia, Punta Arenas (Photo by Michael Pearson 2017).

A number of expeditions took fur seals on the South Shetland Islands later in the nineteenth century. A handful of vessels visited the South Shetlands in the 1830s and 40s as part of wider-ranging sealing and whaling ventures, and ten vessels from Mystic, Stonington and other New England ports made 15 voyages to the islands in the 1850s. Eleven vessels, most of them from New London, operated in the South Shetlands in the 1870s, some of them for multiple seasons, and in the order of 40,000 skins were taken at that time (Headland 2009). Several Nova Scotian sealing vessels operated in the islands throughout the first decade of the twentieth century. It is likely that seal skins gathered after the 1820s were added into a much larger generalised market for seal skins and other furs, as they were a very minor contribution to that larger pool of animal fur (Bob Burton in South Georgia Association 2021).

Sealers were primarily sailors, ranging in experience from able seamen and officers to green-hand first-timers, who happened to get involved in the sealing trade in dangerous and inhospitable environments. They could just as easily have ended up as whalers, or sandalwood collectors in the Pacific, depending on where their ship took them. Companies usually enlisted a significant number of green-hands, as they were paid far less and were considered more obedient and less inclined to desertion (Salerno 2011). In the southern oceans, sealers lived in wet and cold conditions for months on end. Their clothing was that common to seamen, not of the later polar explorer, and comforts were very sparse.

Sealing in the South Shetland Islands was supported by vessels of three types: the largest, the 'mother ship', being the ship, brig, snow or schooner that brought the sealers from their home port and returned them there; the 'shallop', a smaller shallow-draught schooner, cutter or sloop acting as a tender that either accompanied the mother ship from home port or was carried in frame and constructed at a safe harbour, such as in the Malvinas/Falkland Islands or Kerguelen; and boats, ranging from light double-ended whaleboats to ship's skiffs. In the South Shetlands, the threat of uncharted shallows or rocks meant that the larger ships were not able to approach the shore safely but had to find safe anchorages where they moored for the summer. Instead, the sealing gangs, generally consisting of five to eight men, were put ashore in a whaleboat, a lightly built boat of about 9 m long and 2 m wide, with 4–5 pulling oars and a steering oar, often being transported to the beaches by the shallow-draught shallops (see Pearson 2018d for more detailed description).

The overturning of boats in the surf was not unusual and sometimes resulted in loss of life. The boats had little space for cargo, so only essential food with basic cooking equipment and maybe a small amount of coal for fuel was landed. The landing of building materials other than, perhaps, a roll of canvas and some nails was rare. Once ashore, the first task was to find or make a shelter. Most commonly, where seal rookeries were on small beaches, only a day or two's work would clear the beach of seals. So living under the up-turned boat or occupying a convenient cave was much more time-efficient than trying to improvise other forms of shelter (see Pearson 2018a). If the crew had to live in one spot for any length of time, then a shelter had to be built. The evidence of how the shelters were made and occupied comes almost entirely from the archaeological evidence, as described elsewhere in this chapter and the next. In summary, the range of shelters included:

1 two or three curved or straight walls of stone (beach boulders and rocks fallen from cliffs and stacks) projecting from a rock face of a cliff or sea-stack, forming a roughly square or rectangular space, some walls surviving up to about 1 m in height. Archaeological evidence shows that shelters were roofed over using lengths of drift wood (including flotsam and jetsam), whale ribs or jaws, or timber brought ashore, and covered with seal skins or tarpaulins of sail-cloth. In most cases, a gap indicates a doorway, and common features include evidence of fireplaces and whale vertebrae utilised as seats and cutting blocks.
2 four free-standing walls of stone on an open beach, otherwise similar to type 1.
3 occupied cave sites, some with internal stone walls to provide shelter from prevailing winds, with fireplaces and arrangements of stones or whale vertebrae sometimes denoting sitting and sleeping areas.

3.1.2 Early archaeological work in the South Shetlands

Many sealing sites on Livingston Island were identified by Falkland Island Dependencies Survey (FIDS) geologists in 1957–58 and later by British Antarctic Survey (BAS) scientists (Simpson 1959; O'Gorman 1963; White 1967; Hobbs 1968). Lewis-Smith and Simpson subsequently described 15 of these sites on the Byers Peninsula and one on Cape Shirreff (Lewis-Smith and Simpson 1987). Some of the sites were 'excavated' by non-archaeologists as early as the 1950s era ('more of a scrabble really' – Fergus O'Gorman pers. comm., March 2006). This work clearly identified the richness of the archaeological remains that exist in the sealing sites on the South Shetland Islands but lacked any systematic survey or detailed study of the evidence. The sites have subsequently been surveyed and studied by archaeologists, and many more have been found. Lewis-Smith later described a large site at Yankee Harbour on Greenwich Island that might be related to the sealing era (Lewis-Smith 2018; see table of sites for this and other locations in the Appendix).

The first professional archaeological work undertaken in Antarctica was by Rubén Stehberg, an archaeologist with the Chilean National Museum of Natural History, in the South Shetland Islands from 1983 onwards (Stehberg 1983a, 1983b, 2003, 2004). The initial work (1983–95) was directed at resolving the origins of indigenous American artefacts and skeletal material found on beaches. Two possible projectile points had been found in shallow-water samples at Admiralty Bay, King George Island, in the early 1970s (Stehberg 1997), and the skull and femur bones of what was found to be an indigenous woman of about 20 years old were found on Cape Shirreff on Livingston Island (Torres 1992, 1999; Constantinescu and Torres 1995; Stehberg and Lucero 1996). Some of the lithic evidence was found to have been maliciously placed among retrieved marine samples rather than originating in an Antarctica context (Stehberg 1983b), but other evidence pointed to the involvement of indigenous American people as part of the 1820s sealing visits, including the Cape Shirreff skeletal remains and four stone artefacts excavated in a sealing context at Cora Cove on Desolation Island (Stehberg and Cabeza 1984, 1987; Stehberg and Lucero 1985a, 1985b, 1995; Lucero and Stehberg 1996; Stehberg 2003). Excavations were also carried out at Cuatro Pircas sealing sites on the Fildes Peninsula, on King George Island (Stehberg and Cabeza 1984) (Figure 3.3). From 2000, the emphasis of the archaeological work changed to making a systematic inventory of cultural sites in the islands for research, conservation and management purposes (Stehberg 2004: 70). This included further systematic surveys of the Fildes Peninsula that identified much cultural material and nine potential sealing campsites, with minor excavations to clarify the nature of the remains (Stehberg and Gatica 2001; Stehberg 2002, 2003). A site was also excavated at Siddons Point (Punta Negra) on the north coast of Livingston Island (Stehberg 2003: 102–106).

Subsequent work by Stehberg included the survey of Rugged Island and the excavation of two sites there (Stehberg, Pearson and Gatica 2005; Pearson and Stehberg 2006). Rugged Island 1 was an undisturbed cave site (Figure 3.4), with

FIGURE 3.3 Cuatro Pircas shelter sites, Fildes Peninsula, King George Island. Four stone structures on an open beach surveyed and excavated by Rubén Stehberg in 1984 (Photo by Michael Pearson 2007).

a 4 m long timber seal lance with a head of rolled iron lying in the adjacent rocks, fur seal moccasin and shoe parts lying on the cave floor, and a fireplace that was excavated containing artefacts including ceramics datable to the early 1820s (Pearson and Stehberg 2006: 241–243). Rugged Island 2 was a stone shelter against a rock stack, with associated work areas around it (Figure 3.5). The parts

FIGURE 3.4 Rugged Island 1. A cave site occupied by sealers dating to the early 1820s, undisturbed until excavated by Rubén Stehberg in 2005 (Photo by Michael Pearson 2005).

of a dismantled cast iron stove were found in and next to the shelter, which together with copper percussion caps for a firearm suggested the site dated to the mid- to late nineteenth century. The site was initially identified as possibly linked to the second over-wintering in Antarctica in 1872, but the documentary reports supporting this hypothesis have been called into doubt, and the location

FIGURE 3.5 Rugged Island 2. Stone shelter between rock stacks, excavated by Rubén Stehberg in 2005. Occupied in the mid-late nineteenth century (Photo by Michael Pearson 2005).

of that wintering is more likely to have been at Potter Cove on King George Island (Pearson and Stehberg 2006: 344–345; Headland 2018b).

Further work was carried out on Byers Peninsula on Livingston Island, with the excavation of a sealer's sledge near Punta Diablo on the west coast. Timber analysis, thermoluminescence dating of a stoneware bottle and historical context suggest that the sledge was used by a US or Canadian sealing expedition in the period 1871–1908 (Pearson et al. 2008; Stehberg et al. 2009).

The sites found by Stehberg consisted most commonly of collapsed shelters with stone walls built against rocky headlands or free-standing rock stacks. The enclosed space within these structures varied from as small as 3 × 2 m up to 7 × 2 m, and contained cultural deposits with fragments of ceramics, glass, bone, leather and wood. Other occupation sites were free-standing four-walled structures on open beaches, or within overhanging rock shelters or caves. Whale vertebrae, whale ribs and jaw bones were found in many sites. The sites were often well located to provide good visual coverage of beaches and bays and sometimes were located on key paths linking beaches. Other sites consisted of concentrations of timber planks, suggesting timber floors or wall elements of collapsed structures. This pattern has been repeated in other surveys in the South Shetlands, as will be seen below. The factors influencing the location, distribution and construction of sealer shelters were studied by Pearson and Stehberg (2011).

In 1994 a group of geologists from the Argentine Antarctic Institute (IAA), led by Claudio Parica, found cultural remains inside a cave they called Lima-Lima, on the north coast of Byers Peninsula, Livingston Island. The cave had been previously reported in the 1960s (Lewis-Smith and Simpson 1987: 60). A sample of those remains was taken to Buenos Aires, and Daniel Schávelzon (an architect who played a pioneering work in the development of historical archaeology in Argentina) was commissioned for their study. Considering the features of the materials, Schávelzon understood that they could have been the result of the occupation of Lima-Lima by a group of castaways. At that time, it was even speculated that those people could have been the survivors of the shipwreck of the *San Telmo*. With the aim of broadening the study of the site and determining whether it was effectively a shipwreck camp, the IAA asked Andrés Zarankin – an archaeologist from the National Council for Scientific and Technical Research (CONICET, Argentina) – to conduct an on-site study. This work was developed in collaboration with archaeologist María Ximena Senatore (from the same institution).

In the summer of 1995–96, Zarankin and Senatore carried out fieldwork activities for three months on Livingston Island, together with the geologists that had previously identified the site. In addition to studying Lima-Lima, the archaeologists had a chance to visit some of the beaches on Byers Peninsula, identifying multiple sites. The field report made it clear that the occupation of Lima-Lima did not correspond to a survivor camp, but it was part of a broader strategy of exploitation of the South Shetland Islands carried out by sealers. The coastal orientation of the sites, as well as similarities found in the material remains detected there, seemed to suggest this idea. In view of the results of the first fieldwork season, the IAA found it relevant to follow up the research, and a project was created to discuss the early occupation of Antarctica, focusing on the role played by sealers.

After the initial approach, the Argentine project carried out more fieldwork in the summers of 1996–97 (by Zarankin), 1997–98 (by Senatore) and 1999–2000 (by Zarankin). This virtually allowed surveying all the beaches on the north, west and south coasts of Byers Peninsula, and visiting specific locations on Half Moon and Greenwich islands. At that time, a total of 27 sites were identified on Byers Peninsula. Archaeological work generally involved the survey of structures and other remains found on the surface, as well as the use of test-pits. Excavations were restricted to Lima-Lima (where archaeologists established a grid system over the site and excavated a few squares), and Playa Sur 1 and Cerro Negro (which were more intensely excavated).

Analysis of the associated material remains found at the sites was initially conducted to gather techno-functional and chronological information. In the 2000s, several students in archaeology such as Melisa A. Salerno and María Jimena Cruz from the University of Buenos Aires (UBA, Argentina) joined the research project, studying the remains with a view to discuss sealers' practices and experiences in their BA and PhD dissertations (see Chapter 4 for more information). As will be seen in the following section, the 'Landscapes in White' project is a continuation of the Argentine proposal, showing a stronger emphasis on international collaboration. As a matter of fact, some of its researchers are the same, and some of the ideas previously outlined by the Argentine project have been further developed.

Among many other things, the Argentine project stressed the need to understand the early occupation of Antarctica in terms of the global expansion of capitalism and modernity, establishing a dialogue with the new understandings of historical archaeology that gained relevance in the 1990s. From that standpoint, Zarankin and Senatore discussed which aspects of sealers' presence on the South Shetland Islands could have responded to the ways capitalist practices usually expressed themselves in other world contexts, and which aspects were not expressed and why. While the archaeologists understood that sealing companies (represented by the figure of captains on the South Shetland Islands) defined some of the general aspects of the exploitation strategy on the archipelago, including the areas to be exploited and the distribution of sealing gangs on the beaches, several aspects of sealers' lives on the islands were left to their own devices, ranging from the construction of shelters to the gathering of food – see Chapter 4).

3.1.3 Maritime archaeology in the region

In the 1990s, a survey of the remains of the *San Telmo* brought the subject of shipwreck archaeology to the South Shetland Islands. As previously mentioned in the book, the crew of this Spanish frigate, which was last seen partly disabled in the Drake Passage on September 1819, were, if any survived the wreck, possibly the first humans to sight or land on the archipelago. Some spars and an anchor stock were found by sealers on a beach on Shirreff Cove, King George Island, in 1820, raising the possibility that survivors may have been washed ashore up to a month before William Smith made the first recorded landing on King George Island. As indicated below, this has since been confused with Cape Shirreff on Livingston Island. In 1993–94, a Spanish/Chilean team made magnetometer transects across the beach and in shallow waters at Cape Shirreff on Livingston Island but failed to locate any definitive evidence of the wreck site. Some years later the Spanish team returned, this time with Uruguayan archaeologists, in a second unsuccessful attempt to locate evidence of the wreck (Martín-Bueno 1995a, 1995b, 1996; Cabrera 2002; Martín-Cancela 2018). The suggestion made in the listing in 2021 of the yet-to-be-identified *San Telmo* wreck on the Antarctic Historic Sites and Monuments (HSM) list that evidence of shelters occupied by survivors was found by the archaeological surveys seems to be without support.

The location where the possible *San Telmo* remains were found on Livingston Island has since been convincingly questioned. The finder of these ship components, William Smith (and all subsequent reports simply drew on and elaborated on his initial report) in fact found them at 'Shirreff Cove', which he named and landed on, close to North Forehead on the north-east coast of King George Island. This was subsequently confused by later sealers (and historians) with Cape Shirreff on Livingston Island, named by Edward Bransfield, and the original Shirreff Cove and possible wreck site disappeared from the record (Cazenave de la Roche 2019).

Some years after the first survey for the *San Telmo* took place, another collection of shipwreck remains of a potentially identifiable ship has been found on the west coast of Elephant Island in the South Shetlands. In 1998, the United Kingdom's Ice Patrol Vessel, HMS *Endurance* surveyed an area of wreckage of a large wooden sailing vessel in a small cove on the south-west coast of Elephant Island, and timber sample analysis indicated that the timber was not of Scandinavian origin, therefore not a British vessel. Preliminary assessment suggested that the wreckage may be that of the sealing ship *Charles Shearer* that sailed from Stonington, USA, on July 3, 1877, under the command of James Appleton (or Appleman) bound for Antarctica. The ship left a sealing gang on Diego Ramírez Islands in October 1877 and then continued south towards the South Shetlands but disappeared without a trace. The sealing gang on Diego Ramírez was rescued six months later. During the austral summer of 1879–80, two sealing voyages from Stonington searched, unsuccessfully, for the *Charles Shearer* and any survivors. There is a strong possibility that the wreck located on Elephant Island is that of the *Charles Shearer* (ATCM XXII, WP21). This site, if that of the *Charles Shearer,* is the only nineteenth-century sealing site yet recognised on the Antarctic HSM list.

The other shipwreck listed on the HSM list is that of Shackleton's *Endurance,* sunk by ice in 1915. This wreck is usually considered a symbol of the adversities faced by the crew members and their determination to survive, being connected with the drift of the men on a pack of ice, the journey of Shackleton to South Georgia and his return to relief the crew on Elephant Island. In 2019, the wreck was given protection under the HSM list (HSM 93) in view of commercial efforts to find the remains. In that year, an expedition failed due to the loss of their autonomous underwater vehicle in thick ice conditions (Dowdeswell et al. 2020). Another expedition was carried out in 2022 locating and photographing the largely intact wreck. As no physical intervention with the wreck was sanctioned, works involved the recording of the site using an autonomous underwater vehicle. The vessel was found 6 km away from the position recorded by the ship's captain in 1915, and 3000 m below the surface, as this book was being finalised. High-quality photographs of the vessel were taken, and LIDAR recording was employed for the future creation of a 3D model. While this initial precautionary approach has been welcomed, the International Polar Heritage Committee and the International Committee on the Underwater Cultural Heritage have expressed their concern about the future protection and management of the wreck site, a highly desirable target for the collectors of Shackleton memorabilia. The UK is developing a management plan for the site at the time of this book's publication.

The survey for the *San Telmo*, the shipwreck remains on Elephant Island and the finding of the *Endurance* point to the possibilities and challenges for an archaeology of wrecks and campsites (Pearson 2020). Broadly speaking, little research has been carried out in Antarctica or the sub-Antarctic to date on the subject. The international research on shipwreck survivor camps was referred to in Chapter 2, as well as exceptional works at the *Solglimt* wreck and campsite in the Crozets, and campsite identification on Auckland Island. On the South Shetland Islands, a number of shipwrecks and associated campsites on land are known to have existed but have yet to be studied.

The most promising site is that of the wrecks of the brig *Lady Troubridge* (Capt. Richard Sherratt, Liverpool) and the ship *Hannah* (Capt. James Johnson, Liverpool, 320 tons), both sealing vessels blown ashore at Wreck Beach in Destruction Bay, north of Cape Melville, on the north-east coast of King George Island on December 25, 1820. Nathaniel Ames in the brig *Esther* of Boston visited the wrecks in early 1821:

> … we found part of the crew of the two above mentioned ships, engaged in building a small schooner of about fifteen tons, which they called the "*Sailors' Relief*", in which they intended to go to Rio de la Plata. … The *Lady Trowbridge*, a stout new vessel, was apparently uninjured, and at low water we could walk round her; but the *Hannah* had gone to pieces immediately.
> *(Ames 1830: 155)*

The crews of the two ships were picked up by John Davis in the American sealing schooner *Huron* and taken to British ships moored at Rugged Island (Stackpole 1955). A number of locations were also mentioned in the literature as known mooring sites for sealers, such as Esther Harbour (King George Island), Clothier Harbour (Robert Island), Potter's Cove (King George Island) and Yankee Harbour (Greenwich Island). The last two have been surveyed and relatively little land-based infrastructure found (limited to a try-works and possibly an enclosed area at Yankee Harbour –Lewis-Smith 2018: 151–158; Senatore 2018). Esther Harbour is known to be the site of the first wintering of humans in Antarctica in 1821 by a group of men from the British sealing vessel *Lord Melville* (Headland 2018b). Clothier Harbour is the site of the sinking of the American ship *Clothier* in 1820 and the British brig *Robert* in 1822 (Fildes 1821b; Stackpole 1955: 59; Headland 2009: 136). Other than where mentioned, none of these sites has yet been surveyed for archaeological evidence.

An analysis of the timber remains of what appears to be a shipwreck located on the shore near the Uruguayan Station of Artigas on King George Island was begun in 2021 by the Laboratory of Landscape Archaeology and Heritage at the University of the Republic (Udelar, Uruguay). As yet no findings have been made, but one possibility seems to be that the wreckage is material washed ashore from an as-yet unlocated wreck in the adjacent bay (Figure 3.6).

As has been noted, a study has also been carried out on a related maritime archaeological topic – the use of up-turned whaleboats as shelter. This was a common practice in both the sub-Antarctic islands and the South Shetlands, in situations where there were small beaches separated by glaciers or cliffs making the construction of a shelter ashore unnecessary or too time-consuming, or where seals had become scarce and hunting meant rapid movement along

FIGURE 3.6 Wreckage of a supply boat or barge to service whaling ships in the early twentieth century, near Artigas Station on King George Island (Photo by Michael Pearson 2005).

beaches to find them. In such situations, the whaleboat was pulled ashore and one side propped up with stones, whale bones or tussock (in the sub-Antarctic), leaving a doorway to crawl in and out, the men sleeping inside and cooking on a fireplace outside. The hypothesised archaeological footprints of such boat-camps have been reconstructed, including their location above high-water level, and the presence of whale vertebrae and accumulations of stones to hold up the gunwale, or stone alignments placed along the line of the gunwale as a wall or to hold a curtain of canvass over 8 to 9 meters). Several possible archaeological sites of such camps have been located in the South Shetlands, such as Stackpole 1, Stackpole 2 and Punta Vietor 3 on the Southern Beaches of Byers Peninsula, though further archaeological work is still necessary to confirm these interpretations (Pearson 2018a).

3.2 The South Shetlands sealing project – 'Landscapes in White'

As previously noted, little has been written about the lives of people who landed and worked in Antarctica in the early nineteenth century. It is a reasonable assumption that the laws of economics would dictate that the firms that financed sealing expeditions kept their investment in physical and human resources to the minimum necessary to ensure a successful income. Sealing in this context was an enterprise that had the potential for high rewards from modest investment; though it also carried substantial risks associated with long-distance ship operations into dangerous waters. The sealing companies had a vested interest in the survival of its workforce to the extent that its own profits rested on their labour, but survival and comfort were not synonymous and the sealer's life was hard. While the literature to some extent attests to this, especially in the case of the few first-hand sealer accounts that were written, it is the physical evidence found at sealers' camps that really demonstrates the Spartan living conditions in which these people had to endure in the Antarctic environment.

The secondary histories of sealing focus on the economics and mechanisms of the hunt (e.g., Stackpole 1955; Richards 2003), rather than the daily lives of the men involved. Even where logbooks have been published in the past, they have been edited in a way that often excludes references to daily life, social interactions of the sealers or details such as the number of landings effected by the vessels, and the size and composition of the gangs landed at different points of the beaches, although this information exists in some instances in the original texts. Unpublished logbooks are a source worth approaching, as they may contain information not always considered by researchers (see Stackpole 1955 for a pioneering study of these sources). Notwithstanding, recent works have suggested that in order to make sense of the often fragmentary references that logbooks provide on some aspects of sealers' lives on the islands it would be important to go through multiples sources over time, identifying instances of homogeneity and heterogeneity (Salerno and Cruz 2019). This approach would provide relevant information to promote a stronger dialogue with archaeological evidence.

We at times refer to the exploration and exploitation by sealers of a 'territory'. Territory is a theoretical and methodological concept that explains the spatial relationships that human beings establish with certain spaces. It has an empirical context (the area occupied and exploited) but is also a theoretical concept with different meanings in varying cultural contexts. We do not, for example, use it in its political/colonial meaning of a 'claimed territory' – a concept of ongoing relevance to Antarctica as a whole but of little relevance archaeologically. For us, Antarctica as a territory or as a series of territories offers the possibility to understand the multiple changes that could have defined the relationship between humans, and the environment and endemic life systems that exist in this place. Analysing the occupation and exploitation of the Antarctic territory and their changes over time is a central line of study for archaeological work.

Archaeology has the potential not only to fill gaps left in the written records but also to suggest a new approach to the study of the processes of arrival at, occupation and exploitation of this continent. As described above, professional archaeological investigations in the South Shetland Islands began in the 1980s with Chilean archaeologist Rubén Stehberg and a few others in the following years (Martín-Bueno 1995a, 1995b, 1996; Cabrera 2002). In the 1990s, the Argentine team (Zarankin and Senatore 2007) joined the study of this problem, generating new information which expanded knowledge of the scale of sealers' activities in different locations. Archaeology has expanded the knowledge of the human interaction with Antarctica beyond that represented in the written sources, gradually filling the South Shetlands with stories based on the physical record.

This archaeological work has established the sealing history and heritage of areas on Livingston Island, such as Cape Shirreff (Stehberg and Nilo 1983; Stehberg and Lucero 1985a, 1985b) and the Byers Peninsula (Zarankin and Senatore 1999, 2000, 2007); King George Island (Stehberg and Cabeza 1987; Lucero and Stehberg 1996; Stehberg 2003); and Rugged Island (Pearson and Stehberg, 2006). From the surveys and excavations carried out in different sites, they were determined to have functioned as either shelters for human occupation or in relation to the production and storage of seal skins or oil. The spatial and functional organisation of the different camps presents a marked diversity, despite being linked to each other (Zarankin and Senatore 1999, 2007). In general, the date range for the manufacture of the associated artefacts extended from the late eighteenth century into the early nineteenth century, as one would expect in sites of the early 1820s (see Moreno 2000). Artefacts probably dating from the late nineteenth century were reported in a few cases (such as the materials found at Rugged Island 2–see above). Remains frequently reported included clothing, work tools and domestic artefacts through which the sealer's daily lives could be reconstructed. The use of local raw materials was identified both for shelter building and the manufacture of various objects (Senatore and Zarankin 1997).

These studies were characterised by the open exchange of information, and collaborative research programs involving fieldwork, historical studies, laboratory analysis and access to the archaeological collections, publications and management proposals (Stehberg 2004, Stehberg et al. 2008, Pearson et al. 2010, Zarankin et al. 2011). These interactions served as a starting point for the development of the 'Landscapes in White' research project. In 2007, Andrés Zarankin moved from Buenos Aires (Argentina) to Belo Horizonte (Brazil), when he got a full-time position as professor at the University of Minas Gerais (UFMG). There he promoted the establishment of the Laboratory of Antarctic Studies in Human Sciences (LEACH) and the creation of a local group of archaeologists within the structure of the Laboratory to continue the research he had initiated in Argentina. The 'Landscapes in White' project was officially founded in 2009 with the logistical support of the Antarctic Brazilian Program (PROANTAR) and the funding of the National Council of Scientific and Technical Development (CNPq, Brazil).

Right from the start, 'Landscapes in White' maintained close ties with Argentine researchers, adding soon the collaboration of Chilean archaeologists. Some years later, researchers from the United States, England, Australia and Spain also joined the project. This international approach is serving as a vehicle for the integration of all previously gathered information with the production of new data to enable a holistic perspective on sealers' presence in Antarctica. A map showing the location of all sealers' sites so far identified has been prepared as part of the joint effort of archaeologists from Brazil, Chile, Argentina and Australia, and a shared inventory of the archaeological remains collected by Brazil, Argentina and Chile is under construction. This is thought of as a means to transcend national boundaries and produce integral (non-fragmented) knowledge on the region (Zarankin et al. 2011). A number of fieldwork seasons – carried out within the framework of the project – were spent on the South Shetland Islands between 2009 and 2019.

The map in Figure 3.7 shows the magnitude of sealers' presence on the South Shetland Islands. To date, 53 sealing sites have been identified on the archipelago through archaeological surveys – plus one site on Byers Peninsula (Sealer Cave) being recognised by geologists in the 1950s and its material traces no longer found by archaeologists in recent years. Of these sites, 22 sites have been excavated. The Appendix provides a list of these sites, with information on their location, some of its main features and bibliographical references on the archaeological works carried out there. The main places with archaeological sites are on King George and Livingston islands. A total of 8 sites have been reported on the East of King George (with 5 on Fildes Peninsula); and 38 on Livingston (with 28 on Byers Peninsula, 5 on Elephant Point, 4 on Cape Shirreff and 1 on Siddon's Point). As can be observed, Byers Peninsula has the highest density of

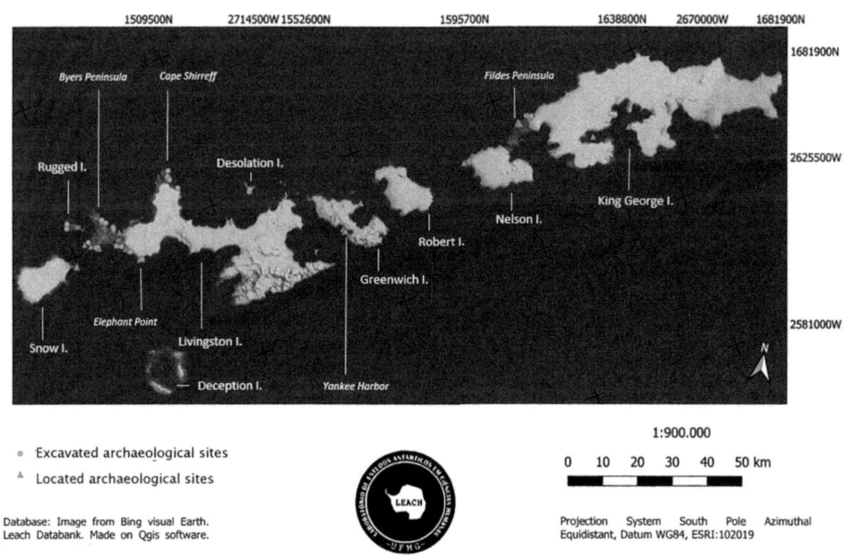

FIGURE 3.7 Map showing the distribution of known sealing sites in the South Shetland Islands (Photo by LEACH 2022).

archaeological sites in the archipelago and indeed in all of Antarctica. A more detailed map of this area is provided in Figure 3.8. It is worth mentioning that, given the logistical difficulties in accessing parts of the archipelago, while most of the likely sealing areas have been surveyed over time, there are still some remote known areas of sealing activity yet to be studied.

Besides the map showing the location of sealers' sites, a dynamic and interactive digital archaeological map is currently being produced through the South Shetland Archaeological Map Project, which allows images (photographs and site plans) and textual information (conservation information, type of archaeological work carried out, reports and publications, etc.), together with videos and virtual reconstructions of the sites, to be accessible by clicking on the map within an interactive information platform. This platform is compatible with Google Earth to allow the public to experience an 'interactive' Antarctica. Associated with the map, another complementary database is being developed with information on the location of the sites, historical data, publications, etc. This information can be accessed on the LEACH website at www.leach.ufmg.br.

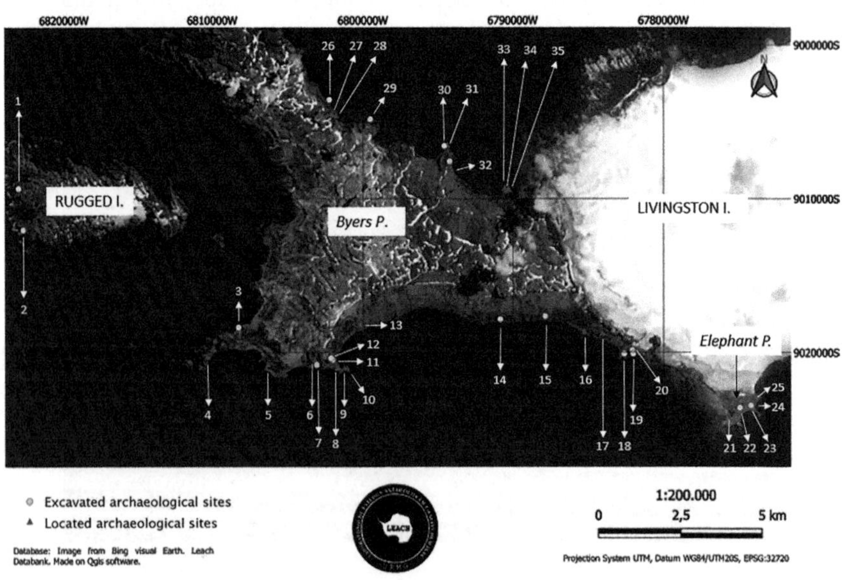

FIGURE 3.8 Map showing archaeological sealing sites on Byers Peninsula, Elephant Point (both on Livingston Island), and Rugged Island (Photo by LEACH 2022). While the Spanish name of the sites on Byers Peninsula and Elephant Point will be used throughout this chapter and elsewhere in the book (considering that most of these sites were identified by Spanish-speaking archaeologists), translations into Portuguese and English are provided as they can also be found in the literature.

Ref. no.	*Site name in Spanish*	*Site name in Portuguese*	*Site name in English*
1	Isla Rugosa 1	Ilha Rugosa 1	Rugged Island 1
2	Isla Rugosa 2	Ilha Rugosa 2	Rugged Island 2
3	Punta Diablo 2	Ponta Diablo 2	Devils Point 2
4	Punta Diablo 1	Ponta Diablo 1	Devils Point 1
5	Rocas Largas	Rochas Largas	Long Rocks
6	Sealer 2	Sealer 2	Sealer 2
7	Sealer 1	Sealer 1	Sealer 1
8	Punta Vietor 3	Ponta Vietor 3	Vietor Point 3
9	Punta Vietor 2	Ponta Vietor 2	Vietor Point 2
10	Punta Vietor 1	Ponta Vietor 2	Vietor Point 1
11	Sealer 3	Sealer 3	Sealer 3
12	Sealer 4	Sealer 4	Sealer 4
13	Playa Sur 2	Praia Sul 2	South Beach 2
14	Cerro Negro	Cerro Negro	Black Hill
15	Playa Sur 1	Praia Sul 1	South Beach 1
16	Stackpole 1	Stackpole 1	Stackpole 1
17	Stackpole 2	Stackpole 2	Stackpole 2
18	Punta X 1	Ponta X 1	X Point 1
19	Punta X 2	Ponta X 2	X Point 2
20	Punta X 3	Ponta X 3	X Point 3
21	Punta Elefante 4	Ponta Elefante 1	Elephant Point 4
22	Punta Elefante 2	Ponta Elefante 2	Elephant Point 2
23	Punta Elefante X1	Ponta Elefante 3	Elephant Point X1
24	Punta Elefante 3	Ponta Elefante 2	Elephant Point 3
25	Punta Elefante 1	Ponta Elefante 1	Elephant Point 1
26	Pencas 3	Pencas 3	Pencas 3
27	Pencas 2	Pencas 2	Pencas 2
28	Pencas 1	Pencas 1	Pencas 1
29	Punta Varadero	Ponta Varadero	Varadero Point
30	Lima-Lima	Lima-Lima	Lima-Lima
31	Punta Lair 1	Punta Lair 1	Lair Point 1
32	Playa de los Ladrones	Praia dos Ladrões	Robbery Beach
33	Cutler 2	Cutler 2	Cutler 2
34	Cutler 1	Cutler 1	Cutler 1
35	Cutler 1 – Calderas	Cutler 1 – Caldeiras	Cutler 1 – Try-works

The material remains that are part of the archaeological collection kept by LEACH, in Brazil, have been thoroughly studied attending to technological, morphological, stylistic and conservation features, among others (see Chapter 4 for more information on the analysis of artefacts and other material traces) (Soares 2020). With the aim of simplifying data logging and access, a special database has been created, categorising the main attributes of the artefacts that will enable relevant analysis. This database is currently being employed for documenting other archaeological collections and creating a unified and standardised inventory of materials. A series of guidelines are followed to orient the study, and new adjustments are made to the database in accordance with new insights and recommendations. Information on different objects kept by LEACH is available on its website (www.leach.ufmg.br).

Fieldwork activities by the 'Landscapes in White' project took place in 2010, 2011, 2012, 2014, 2017, 2018 and 2019, counting on the presence of researchers from Brazil and from some other countries. Activities mainly focused on Byers Peninsula and Elephant Point, on Livingston Island. Work on Byers Peninsula allowed expanding archaeological approaches by the Argentine project in the 1990s. This was particularly relevant to monitor the condition of previously identified sites, detect archaeological remains as yet unrecorded, record the coordinates and digitally scan some sealers' camps, and conduct observations and excavations at different places, etc. Unlike Byers Peninsula, Elephant Point had not been previously approached by archaeologists. Therefore, works on the area gave a chance to broaden the archaeological coverage of Livingston Island. Other areas surveyed by the members of the research project included Snow Island and several small islands localised at the north of Greenwich and the west of Robert Island.

Within the 'Landscapes in White' project, two different scales of analysis can be identified: the sealer's camp and the broader sealing landscape. This differentiation is relevant to understand the processes of occupation and exploitation of the region and recognise the existence of similarities and differences in the organisation of sealing activities at the specific shelter level and in the surrounding areas. The identification of sites within landscapes also has implications for the future management of them as Antarctic heritage.

3.2.1 Approaching sealing landscapes and camps

One of the first steps taken by researchers to approach the sealing landscapes is to undertake surveys. The latter are key to identify archaeological sites and describe the spaces between and around them reflecting past human activities – in particular, the activities carried out by sealers while working and living on the islands. Attention is mainly focused on determining the presence/absence, abundance, diversity and distribution of material remains including structures and objects. The impact of geographical, functional and social aspects on the use of space is especially considered, including discussion on the influence of sealing on the patterns of human use of the coasts vs. the interior of the islands. As might be expected, archaeological remains found during surveys are clearly scarce away from the coastline, reflecting a primary focus on coastal activities such as taking seals and elephants seals lying on the beaches, or embarking and disembarking people and things.

3.2.2 The sealing process

In order to understand the range of activities in the sealing landscape, and the range of production tools found in archaeological sites, it is necessary to briefly describe the sealing process. More detailed discussion of the artefacts associated with sealing is provided in Chapter 4. The killing and skinning of fur seals was a brutal and bloody process (Nolasco 2018). Techniques seem to have varied considerably: in one account sealers lined up along a beach, then walked inland clubbing seals over the nose as they went, allowing the more aggressive males to escape to the sea (Ames 1830: 145); while in another account, three or four pairs of sealers formed two lanes of men and drove the seals between them, sometimes dividing work between stunning seals by clubbing on the nose and killing them with a knife to the heart (Delano 1817: 306; see also Kirker 1970: 20; Clarke 1850: 25–26).

Sealers had to adapt their approaches to suit the season and the number of seals available. Even at the peak of the sealing boom in the South Shetlands, Nathaniel Ames records that towards the end of the season '… the seals were so very rare and shy, that we had much trouble to get two skins a day among five of us, though we had six miles of beach to travel over and examine every day', and he refers to taking seals individually from isolated rocks along the shore (Ames 1830: 144, 148). While firearms made it possible to kill the animals from a distance, they also caused damage to the skins. Furthermore, as the animals dropped dead, they could fall into the sea or slip between cracks and crevices, making their recovery an additional challenge for the hunters (Clark 1887: 434).

After the slaughter, all hands were employed in skinning and preparing seal skins. In addition to a club, each sealer carried a set of knives in a case attached to his belt (described in detail in Chapter 4). The seal had to be skinned soon after killing, as skinning a stiff carcass was difficult. To remove the skin, a cut was made around the neck, above the line of the ears, and then the skin was cut down from the chest to the lower belly. The flippers were cut around, leaving them on the carcass. Some amount of fat was left on the skin, as it was easier to do this than to risk cutting the skin by getting too close (Delano 1817: 30; Clarke 1850: 26; Phelps 1871: 64; Clark 1887: 433). The skins were beamed, the excess blubber was shaved off with a long blade, the flipper holes were sewn up and the skins were washed to remove blood and sand. The skins were then either salted or dried, depending on the target market: at least into the early nineteenth century China preferring dried skins and the markets in Britain and America preferring them salted (Delano 1817: 306; Clarke 1850: 26–27).

If the pelts were to be salted, a thin layer of blubber was left on the skins, and they were stacked up to 'sweat' for a few days. Then the skins were interlayered with salt and re-stacked in barrels or in bins constructed in the ship's hold, the salting process having to be repeated several times. Sealing ships diverted via the Cape Verde Islands to collect salt from the late eighteenth century onwards, and Cape Verde islanders became an increasing proportion of US ship crews by the 1840s (Delano 1817: 306; Fanning 1833: 358–359; Townsend 1797 in Clark 1887: Goodridge 1839: 29; 435; Phelps 1871: 64; Clarke 1850: 26–27).

The skins for the China market were stripped of attached blubber and then pegged out to dry after the flipper holes had been sewn up. Ten to 14 wooden pegs were used, drying taking one or two days (Moulton 1804; Ames 1830; Goodridge 1839; Clarke 1850; Phelps 1871: 64; Howard 1940: 40; Calder 1984; Nolasco 2018). Delays in pegging or the poor management of skins, salted or dried, could lead to their decay (Moulton 1804: 61; Ames 1830: 160).

Elephant seals were killed by either lancing them in the heart or shooting them in the head, sometimes after hitting them on the nose with a club to stun them or to make them sit up and expose their breasts or open their mouths. The thick blubber was cut off the carcass, which had to be rolled over to complete this process – a difficult task with an animal that weighed up to four tonnes. Trying out was done ashore at try-works with two or more large iron try-pots, or the blubber was rafted aboard the ship and tryed-out there. Several try-works sites, marked by stone circles and a dense residue of burnt blubber, have been located in the South Shetlands at Yankee Harbour (together with a broken try-pot), Flat Top Head at Fildes Peninsula on King George Island, and Cutler Beach at Byers Peninsula on Livingston Island. Large sections of blubber were sliced into smaller portions with a blubber knife on a wooden bench and then fed into the try-pot (Clarke 1854; Phelps 1871; Richards and Knight 1996; Nash 2009).

3.2.3 Sealers' camps – their location and distribution

The South Shetlands Islands were entirely new territory, further south than had ever been occupied before, unknown and with no maps. Getting to know the new territory and its resources was made even more complicated by the fact that it was only accessible by sailing vessels for about three months of the year – the winter off-seasons were often spent in the slightly milder Malvinas/Falkland Islands, over a 1000 km to the north. We know something about how the charting of the South Shetlands unfolded from the maps that were prepared by the British sealers – the American sealers made none (Pearson 2020). The main evidence of where the sealers did their work, however, comes from the physical survey of the sites themselves.

When sealing vessels arrived at the South Shetlands, the two immediate aims of the captain were to find beaches with seals on them and land sealing gangs there, and to find a safe harbour in which to moor the ship. Most ships were left in the harbour for the duration of the season, with upper masts taken down. The daily work of finding beaches and landing and provisioning gangs was done by the shallops. Such was the need for manpower that the ships were often left with just skeleton crews aboard. Robert Fildes left his ship *Cora* moored at Clothier Harbour on Robert Island in December 1820 with two active and two sick men aboard and two men ashore beaming skins (Fildes 1820–21: *Journal of the Cora*, Dec 22, 1820). Fildes subsequently wrote of Clothier Harbour on Robert Island: 'The *Robert* and *Mellona* lay so snug in this place that all hands were away for many days together leaving only a little boy as shipkeeper to take care of the vessels' (Fildes 1821a: 'Remarks made during a voyage to New South Shetland'). The inherent risk inbuilt into this strategy was shown when Fildes' ship *Robert* was subsequently wrecked during a storm at this anchorage while he and nearly all the crew were absent. In another case, Ebenezer Townsend was left alone as shipkeeper on his ship in the Malvinas/Falkland Islands while the shallop went off sealing for two weeks.

> but as it was a good harbor and the ship had good cables and anchors, I felt safe. I was then captain, mate, and all hands. As I had enough to do I was not so lonesome as you may imagine.
>
> *(Townsend in Clark 1887: 462)*

To get gangs to sealing beaches, the shallop was normally used to transport whaleboats and gangs, drop them at beaches and return at intervals to resupply the gangs and take off skins or oil/blubber. When the animals were scarce, the gangs might only have been ashore from a couple of hours to a couple of days at a given location. Sometimes men were sent off in whaleboats to search for seals along a given coastline and re-join the ship at a later time (Pearson 2018a). When the seals were found in large quantities, sealers might spend weeks at a single place. Each of these options had specific consequences on the organisation of sealers' lives on the islands and in the archaeological footprint of their presence.

The sealer shelters, the general nature of which was described earlier, were either stone-walled residential huts, occupied caves, work and storage spaces or small structures, the use of which is still unclear. The shape, size and number of rooms or associated structures found in sealing camps varied. Most sites contained only one or two structures, although others had multiple stone-walled structures. Structures might be square, rectangular or round. Some were small (having enough room for only one person), while others were a little more spacious. In general, none of these structures exceeded 15 m^2 (except for one shelter built inside Lima-Lima Cave). A consideration limiting the size of the structures was the availability of roofing members needed to span the walls: they were either whale ribs or jaws, or timber manhandled ashore from whaleboats, and in both cases three metres or less was the practical limit. Small spaces are also easier to heat than larger huts, and unconsolidated stone walls can only support relatively light-weight roofs.

On the beaches of the South Shetland Islands, the distribution of sealing sites was dispersed and uneven. The pattern of sealing sites appears to reflect a number of factors: the occupation, where feasible, of beaches where seals were seen breeding or congregating; the degree of danger involved in approaching and resupplying the target beaches from the sea; the availability of drinking water and living sites safe from boggy land, storm surge or ice falls; the need to be close to the seal supply and to guard against poaching by other gangs; the vigorous protection of some beaches by sealing gangs from different countries or competing sealing companies; and a lack of incentives for sealers to move away from the coastal strip (Pearson and Stehberg 2011). Distinguishing the functional or operational relationships between adjacent archaeological sites is difficult, as neighbouring sites could have been established independently at different times or could have served complementary functions during the same occupation period.

An important example of this, demonstrating the interrelated role of historical and archaeological evidence in understanding the landscape, is the distribution of American sealers on the Southern Beaches of Byers Peninsula on Livingston Island. In the 1820–21 season, Captains John Davis (of the ship *Huron* and shallop *Cecilia* of New Haven) and Christopher Burdick (of the schooner *Huntress* of Nantucket) joined forces in sealing (Burdick 1821; Davis 1822; Stackpole 1955). Their ships were anchored for the season in Yankee Harbour, Greenwich Island, and the shallop *Cecilia*, a small schooner, explored and serviced the sealing camps on the Byers Peninsula. Davis and Burdick established three camps manned by 22–28 men along the Southern Beaches of Byers Peninsula. From the geographical descriptions and distances recorded in the ships' logs, the three camps correspond to the archaeological sites at Elephant Point (the Camp 1 under First Mate Coleman of the *Huntress* with eight men), Playa Sur or Punta X (Camp 2 under Third Mate Ripley of the *Huron* with possibly seven men) and Sealers Hill/Vietor Point (Camp 3 under Second mate Phillips of *Huron* with possibly seven men). Elephant Point, while separated from the rest of Byers Peninsula by three kilometres of glacier, was regarded as the eastern-most part of the Southern Beaches (described on a map by Robert Fildes, 1821a).

Davis's men reported that there were at least 50 other American sealers working on the Southern Beaches when they arrived, mainly from a fleet of sealing vessels from Stonington, Connecticut, which would account for the other campsites recorded there. The British sealers were gathered on the Northern Beaches of the Peninsula (Stackpole 1955). As indicated above, there were often tensions between sealing gangs established on beaches and newcomers trying to get a foothold. Thomas Smith, an English sealer on the South Shetlands in 1820–21, recalls approaching a new beach:

> on our landing, we were met by the crews of three vessels, who forbid us from taking any, claiming the beach as theirs, as they had first taken possession of it and were therefore determined to defend it as their own ground to the utmost in their power. ... being unable to enforce our rights against superior numbers, we reluctantly left the beach and returned to the vessel....
>
> *(Smith 1844: 159–60)*

Smith later recounts how, having found a beach, he claimed similar ownership rights when others arrived and

> ...attempted a landing on our beach, but having notice of their coming, we concentrated our forces, and prevented their landing in spite of all their efforts to effect it. — Being defeated in their undertaking, they reluctantly returned to their ships, which were making ready to sail for England.
>
> *(Smith 1844: 160)*

Captain Robert Fildes describes a similar confrontation on Livingston Island in December 1820, and personally went to negotiate the sharing of a beach, only to have his ship, the *Cora*, blown ashore and wrecked in his absence (Fildes 1820–21). A potentially major armed confrontation between American and British sealers at Cape Shirreff on Livingston Island in January 1821 was avoided only because it was seen that the seals on the disputed beach had already been killed (Burdick 1821: Jan 25–27, 1821; Stackpole 1955: 44–46). This aspect of sealers' experiences is only available to us through written journals, as it left no material evidence in the archaeological record.

3.2.4 Sealers' camps – their use and organisation

Sealing camps varied in the number, size and shape of their structures. These may have been influenced by the number and composition of the sealing gangs, the length of the time spent ashore and the type of production carried out at a given location (the hunting and processing of either fur seals or elephant seals). Archaeologists have often distinguished between structures used for accommodation and others associated with work. Size has often been associated with the potential use of the structures, the larger ones being probably oriented to subsistence activities and the smaller ones to production. The presence of specific features within the inner space of several large structures, such as hearths and whale vertebrae around them used as furniture, has also been considered suggestive of their use as accommodation shelters.

In spite of this, the key evidence enabling the reconstruction of the life experienced by sealers in the South Shetlands is provided by artefacts and some other remains. Artefacts associated with domestic life, food and cooking remains were primarily found within the larger stone shelters, while working tools were often found in adjacent smaller structures, designated as 'annexes', suggesting the separation of living and work areas at the sites. In the accommodation shelters, material remains are associated with sealers' food, clothing, leisure time and other domestic practices. Artefacts in spaces devoted to work suggest activities such as beaming skins, sharpening tools and certainly for storage, as skins have been found inside some of them. Shelters thought to be devoted to work were often smaller. These structures also contained very few remains of subsistence activities (Figure 3.9).

Even though artefacts and the cultural practices they were connected with will be discussed in more detail in the next chapter, here it is worth presenting some information with a view to understand the organisation of sealing camps. Archaeological evidence for the sealing processes includes wooden pegs used for pegging out skins that have been found in several shelters, suggesting that they were being manufactured ashore. Rings of burnt ash and concreted material have been found indicating the sites of try-works, although except in one case the try-pots, being expensive equipment which needed to be used in some other locations, were removed when sealing ended. Other finds probably associated with

FIGURE 3.9 Artefacts in the archaeological context of sealers' sites on Byers Peninsula, including a shoe, a whale vertebra used as furniture and a pipe (Photo by LEACH 2010).

processing activities include the remains of iron hoops used to fasten wooden casks, timber cask staves and heads, large cutting and flensing knives, lance heads and fragments of whet stones.

Archaeological evidence also reveals aspects of the daily life of sealers on the islands, such as cooking and eating, clothing and recreation (Zarankin and Senatore 2007). Historical documents suggest that sealing gangs often brought some supplies from the vessels. Material evidence indicates the use of coffee, alcoholic beverages and salted pork and beef. The documentary sources and the archaeological evidence also indicate, however, that the sealing gangs had to resort to local resources to survive, the food supplies and particularly the fuel supplies taken ashore not being sufficient to last for long periods away from the ship. The sealers ate parts of the fur seals and elephant seals they hunted for commercial gain (Cruz 2011, 2014), and elephant seal blubber has been found in many fireplace residues having been used as fuel (Villagran et al. 2013: 12). They also ate birds and gathered their eggs. In the sub-Antarctic islands, they ate local vegetation, commonly known as Kerguelen or Macquarie Island cabbage. Clothing, as described in Chapter 4, consisted of woollen, flax or cotton fabrics, and leather shoes, such as those advertised by the outfitters at various ports. Clothing and shoes were in constant need of repair or modification as rough conditions wore them out. The manufacture of shoes and moccasins with skin leather on the islands is attested by documentary and archaeological evidence (Salerno 2006, 2009, 2011; Alkimim 2015).

3.2.5 Sealers' camps – some examples

Described here are a number of sites on the Byers Peninsula and Elephant Point on Livingston Island, an area where the 'Landscapes in White' project has identified the greatest density of archaeological sites in all the South Shetland Islands. The types and functions of the sealers' shelters are illustrated by looking at the case studies of Lima-Lima (on Byers Peninsula, Northern Beach); Punta Elefante 2 and X1 (on Elephant Point); Playa Sur 1, Stackpole 1, 2 and Sealer 1, 2, 3 (all these on Byers Peninsula, Southern Beach).

These case studies show some basic differences in the characteristics of the structures and reflect the variety of strategies that the sealers used to occupy and exploit the territory. Two questions guided the analysis: what can the sites tell about the organisation of space, considering the structures, and the relationship between each other and with the surrounding areas, and what can the sites tell about the duration of residence and the possibility of multiple occupations of shelters over several seasons? In the latter context, it should be noted that while a number of sealing ships returned to the South Shetlands for two or more seasons, the available logs and journals do not indicate whether the sealers utilised the same beaches, or re-occupied the same shelters. The extent to which these questions can be addressed at particular sites is complicated by the generally limited thickness of the occupation layer, which in itself might indicate one-season use of most sites.

- *Lima-Lima*

  62° 36′ 55.62″ S, 61° 02′ 13.08″ W

  Lima-Lima site is located inside a cave (Figure 3.10), which has two branches or lobes and a total area of approximately 22 m deep and 14 m wide. At the inner end of the cave is a small man-made wall of overlapping rocks that delimits an internal area of approximately 22.75 m^2, identified through excavations as the living area (Figure 3.11). The cave is at a distance of 500 m from fresh water and 400 m from the landing beach, but it does not have a clear view over the beaches where the seals gather, so it appears to have been chosen for its quality as a shelter rather than for the other locational attributes we have previously described (especially surveillance). While there is no clear evidence of multiple occupations of the site, the extent of the artefact scatter and its composition suggests a lengthy occupation over at least one season.

  The Lima-Lima site has a history of interventions by British geologists dating back to the 1950s. It was known to them as Lair Cave (see Simpson 1959; White 1967; Lewis-Smith and Simpson 1987). The British visit in 1959 excavated parts of the shelter with no stratigraphic or plotting control, and all the artefacts recovered were subsequently discarded by the Scott Polar Research Institute. Significant artefacts were found, including a wooden knife sheath and skin and canvas moccasins. The stone wall demarking the

FIGURE 3.10 Exterior view of Lima-Lima Cave, Byers Peninsula, Livingston Island (Photo by LEACH 2019).

FIGURE 3.11 Excavations in Lima-Lima, 2018–19, showing the remains of a stone wall across the rear of the cave (Photo by LEACH 2019).

living space supported a number of wooden poles, which are presumed to have carried a cloth protective curtain to give added protection from winds (Lewis-Smith and Simpson 1987: 67, 69).

As previously mentioned, in the 1990s, geologists from the IAA made surface collections and informed archaeologists about the site, and in 1995–96 Andrés Zarankin and María Ximena Senatore carried out systematic and controlled investigations. In 2018 and 2019, the excavations were resumed by the 'Landscapes in White' research project, under the coordination of the LEACH and its director, Andrés Zarankin. A precautionary approach was taken to the investigations, given the historical importance that a site of this size and type has for understanding the presence of sealers in the region. Thermo-luminescence dates obtained by Rubén Stehberg (pers. comm. 2008) gave a date range between 1805 and 1830.

In 2018 and 2019, approximately 60% of the site was excavated using a 1 m^2 grid and with the alphanumeric designation correlating with the earlier work by the Argentine team (Figure 3.12). About 850 archaeological remains were collected, including metal and wooden buttons, wooden pegs,

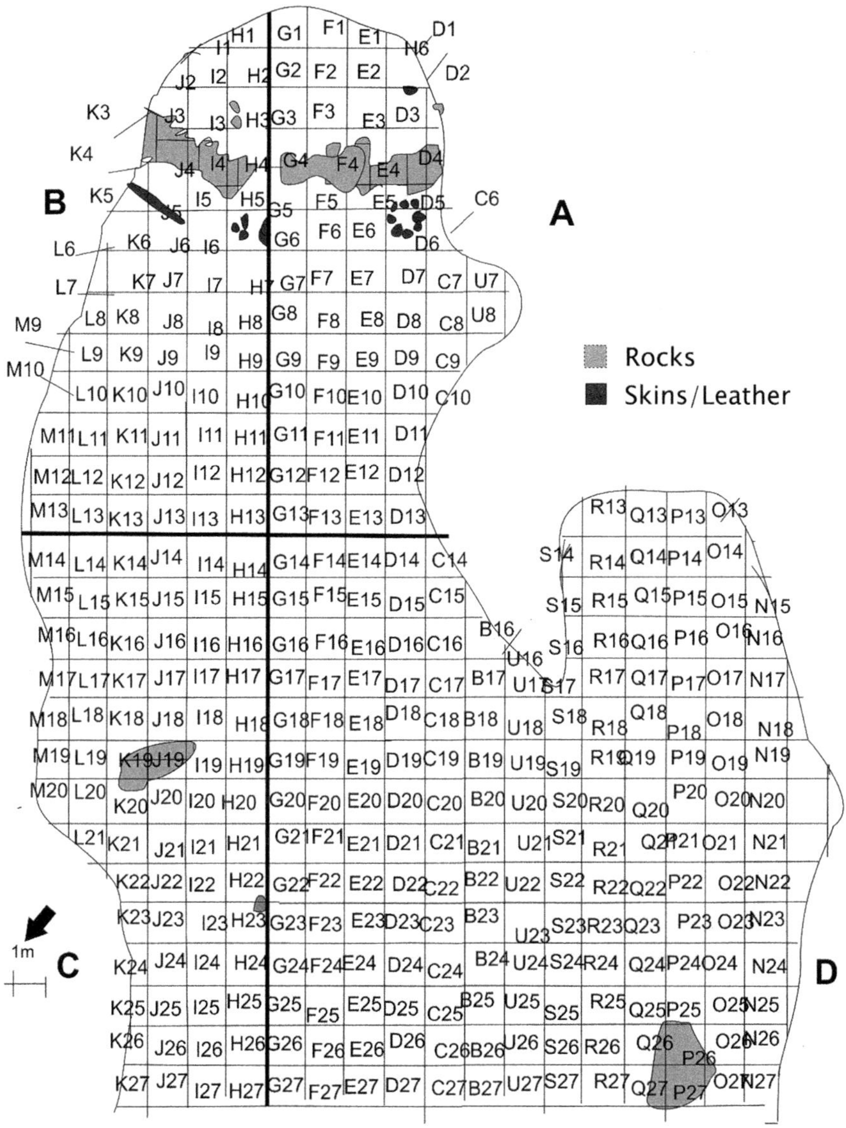

FIGURE 3.12 Site plan for Lima-Lima, showing the grid used by archaeologists (Drawing by LEACH 2019).

leather shoes, fragments of pieces in crockery and stoneware (possibly plates and bottles), fragments of clay pipes, glass bottles, a wood and graphite pencil, metal trouser braces and others. Soil samples were also carried out for studies (Valadares 2020). In all, approximately 50 grid cells were excavated (Figure 3.13). A 3D scan of the cave was also carried out (images of the 3D model produced by the project can be found at https://sketchfab.com/leach.ufmg)

The intervention was carried out through careful stripping and respecting the natural stratigraphic levels of occupation of the site. The archaeological stratigraphy is characterised by the following:

- **Surface level**: It comprises rocks detached from the cave roof, large amount of greenish algae and/or mosses and scattered archaeological materials.
- **Level 1**: This is the body of the archaeological occupation of the site. The soil has a darker colour. A large amount of archaeological remains and some pebbles were found.
- **Level 2**: This is the sterile level of the site, where several stones were identified (similar to those found on the beach) and practically no archaeological remains were found (except for some minimal fragments that migrated from level 1 as a result of formation processes).

The recovery of archaeological material followed the precepts of preventive conservation actions (Figure 3.14), which consisted of any necessary

FIGURE 3.13 Excavations at Lima-Lima, 2018–19 (Photo by LEACH 2019).

triage treatment while in situ and while lifting, and supplementary curative conservation measures in the laboratory set up in the camp. This was followed up with further conservation action as required on return to Brazil.

Lima-Lima is a large site, and contained a lot of archaeological evidence despite the large amount removed and disturbed in the 1950s. The timber poles set upright in the stone wall within the shelter, noted in 1959, indicate considerable efforts to make the cave habitable, and smoking pipe fragments and the large amount of shoe leather and moccasin components uncovered then and subsequently – among other things – indicates ample leisure-time pursuits, suggesting a long occupation. The evidence (especially with the disturbance of stratigraphy by earlier digging) does not allow any conclusion as to whether it was occupied for the large part of one season or was re-occupied over several seasons. The cave is located about 100 m from another occupation shelter (Punta Lair 1), which does provide visual oversight of a section of beach (which Lima-Lima does not), but it is unknown if the two sites were functionally related or occupied instead by two of the many gangs that worked the Northern Beaches in 1820–22.

- *Punta Elefante 2*

  62° 41′ 06.9″ S, 60° 51′ 13.8″ W

  Punta Elefante 2 is arguably the best preserved sealing site among those found on the South Shetland Islands (Figure 3.15). It is made up of a square-shaped enclosure, located in the shelter of a rocky outcrop, enclosed by

FIGURE 3.14 Conservation work being undertaken on artefacts as they are excavated at Lima-Lima (Photo by LEACH 2019).

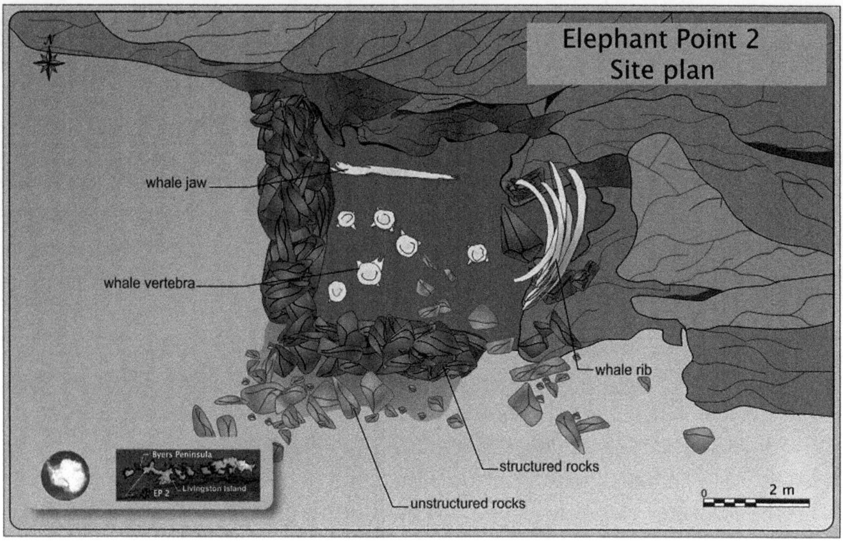

FIGURE 3.15 Site plan for Punta Elefante 2 (Drawing by LEACH 2014).

stacked drystone walls, one of which is practically intact to its apparent original height of 90 cm. The distance between the site and fresh water is 600 m, and the site is 70 m away from the sea and landing beaches. The drystone stone wall is 0.40 m thick and encloses a space of 5.00 m × 3.00 m, with a surface area of 15.00 m^2.

Inside the structure, a collection of five whale ribs are propped up at one end, above the site of the large fireplace. It would appear that the ribs, which are interpreted as being the roof beams, may have been moved to one end when the shelter was abandoned, perhaps to provide easier access to remove the contents or to secure the structure for future use (Figure 3.16). A large whale jaw bone was found on the floor, perhaps part of the roofing system, or a partitioning element within the hut. Whale vertebrae, some with scoring on their upper surfaces, appear to have been used as seats, cutting blocks or tables (or all three). The excellent preservation of the site has to do with its location in a niche of the cliff, which gathers drifts of snow that almost totally covered the site all year round protecting it from animals and winds. The site has therefore been undisturbed, the excavation of the site showing what is thought to be the primary context of occupation.

The analysis of this site suggests that it was worked together with Punta Elefante X1 (which was also excavated), Punta Elefante 2 being interpreted as a place of residence and Punta Elefante X1 as one of work. The historical evidence (the Davis and Burdick journals summarised in Stackpole 1955 and described above) suggest that the site was occupied by an American sealing gang of approximately seven to eight men who were relieved and replaced several times over a period of two to three months. There is no evidence of

FIGURE 3.16 Excavations underway at Punta Elefante 2 shelter site, showing the whale rib roof frames and vertebrae furniture in situ. Stone walls to lower right and left foreground (Photo by LEACH 2014).

multiple occupations, but the arrangement of the roof beams may suggest that the shelter was 'mothballed' when the last crew left in case it needed to be reused in the following season. The Davis and Burdick evidence also suggests that the site might be functionally related to Playa Sur 1 (see below) and the Sealers Hill/Vietor Point sites to the west.

The excavation in 2014 was carried out through careful stripping and respecting the natural stratigraphic levels of occupation of the site. The archaeological stratigraphy is characterised by the following levels:

- **Surface level**: It comprises rocks detached from the cliff, large amount of greenish algae and a thin layer of sand.
- **Level 1**: This is the body of the archaeological occupation of the site. The soil has a darker colour with fragments of burnt bone and contains all of the archaeological remains.
- **Level 2**: It is a permafrost soil, and the sterile level of the site, with no archaeological remains.

Punta Elefante 2 is the only site yet located with major roof components in situ and has a substantial array of apparently undisturbed elements such as the fireplace and whale vertebrae furniture. The site is, on the one hand, typical of many other shelters, but its condition and intactness marks it out as a 'site type' illustrating the key components the shelters would have had when in use. The site is located in easy reach of the main beach on the peninsula but does not have a clear view of the beaches. Currently, Elephant

Point is usually visited by large numbers of tourists. Therefore, it would be relevant to protect the sites from possible disturbance.

- Punta Elefante X1

  62° 41′ 51.8″ S, 60° 51′ 34″ W

  Punta Elefante X1 has a rectangular shape, defined by stacked drystone walls, so that its location allows the use of a rocky outcrop as the back wall (Figure 3.17). The distance between the site and fresh water is 550 m, and it is 30 m away from the sea. The stacked stone wall is 0.60 m thick and encloses a space of 6.00 m × 2.80 m, with a surface area of 16.80 m^2. The site was almost entirely covered with a depth of penguin guano, feathers and beach sand with just the top of a few stones exposed when excavation began.

  As mentioned earlier, Punta Elefante X1 is interpreted as being associated with the Punta Elefante 2 site, which is about 50 m walking distance away, and as being used for some activities of animal processing such as storage and maybe cleaning seal skins. Its proximity to the wide upper-beach berm might also suggest that it was connected with a skin pegging area. Approximately 60% of the site was excavated (Figure 3.18), leaving an intact sector for future work. It is a room without divisions, and a large fireplace was located inside. There is another fire on the outside of the wall against the outcrop. One reason for suggesting that the Punta Elefante 2 and Punta Elefante X1 sites were functionally associated is that Elephant Point has a relatively small beach area (about 1 km comprised of four beaches separated by headlands) and is unlikely to have supported more than one sealing gang

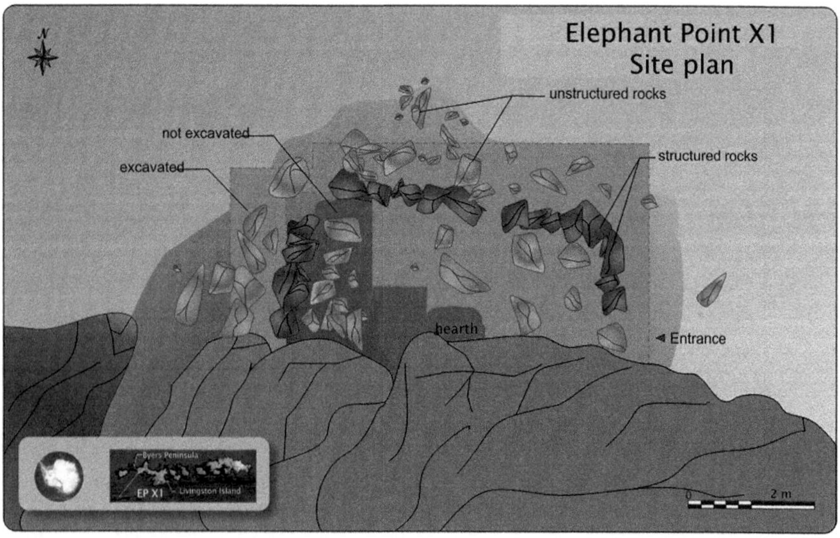

FIGURE 3.17 Site plan for Punta Elefante X1 (Drawing by LEACH 2014).

FIGURE 3.18 Excavation of Punta Elefante X1, 2014, showing the stone walls against the rock outcrop (Photo by LEACH 2014).

at a time, although it is feasible that a second gang could have been located at Punta Elefante 1 site at the western end of the point, about 1 km away from Punta Elefante 2. Given the proximity of two other shelters, Punta Elefante 3 and Punta Elefante 4 to Punta Elefante 2 (100 and 150 m, respectively), it might be interpreted that they, too, were related either to the Punta Elefante 2 operations or to a separate gang in a subsequent season.

The dig in 2014 was carried out through careful stripping and respecting the natural stratigraphic levels of occupation of the site. The archaeological stratigraphy is characterised by the following levels:

- **Surface level**: It comprises rocks detached from the cliff, a large amount of greenish algae and a thin layer of sand.
- **Level 1**: This level is composed of a compact sandy sediment of light colour, in which some dispersed archaeological materials appear. At the base of this level, overlaying the occupation layer below, is a thin black layer, identified as possibly a tephra deposit from an eruption of nearby Deception Island volcano, which might have immediately post-dated the first sealing period (discussion with geologist and volcanologist John Smellie, 2014).
- **Level 2**: This is the level with a dark clayey sediment, in which small fragments of bone appear (common in all the habitation sites we excavated), and it was interpreted as the occupation floor. Most of the archaeological remains found were at this level.
- **Level 3**: It is the last level, sterile, with a mixture of sand with pebbles and no human action.

- *Playa Sur 1*

  62° 39′ 40.02″ S, 60° 58′ 34.56″ W

  Playa Sur 1 is an isolated site 3 km away from the nearest structure along the shore, and it was excavated by the Argentine project in 1995–96 and 1996–97. It is 10 m from fresh water and 25 m from the sea. It is made up of a main shelter enclosure and a smaller one (the 'Annex') (Figure 3.19). The larger structure covers an area of 12.40 m^2 and is square-shaped. It is built with two stacked drystone walls connecting two low natural rock stacks that form the side walls. The stacked stone walls are 1.60 m high and 0.40–1 m thick. Next to the east wall, fur seal skins, iron nails and pieces of rope which were part of the roof were found. The main enclosure provided shelter to a small group of people. It presented no segmentations or divisions with no differential or hierarchical structuring of the use of space.

  Inside the main shelter were small fire-pits and a series of whale vertebrae, interpreted as being furniture. Artefacts and organic remains were concentrated around these features. Archaeological findings indicate the consumption of food, drink, tobacco, the mending and making of clothes and tools (for instance, wooden pegs such as those used for pegging out skins). Most of these remains were found in a large concentration of materials next to the enclosure door, which was interpreted as a dumping area. The pattern and

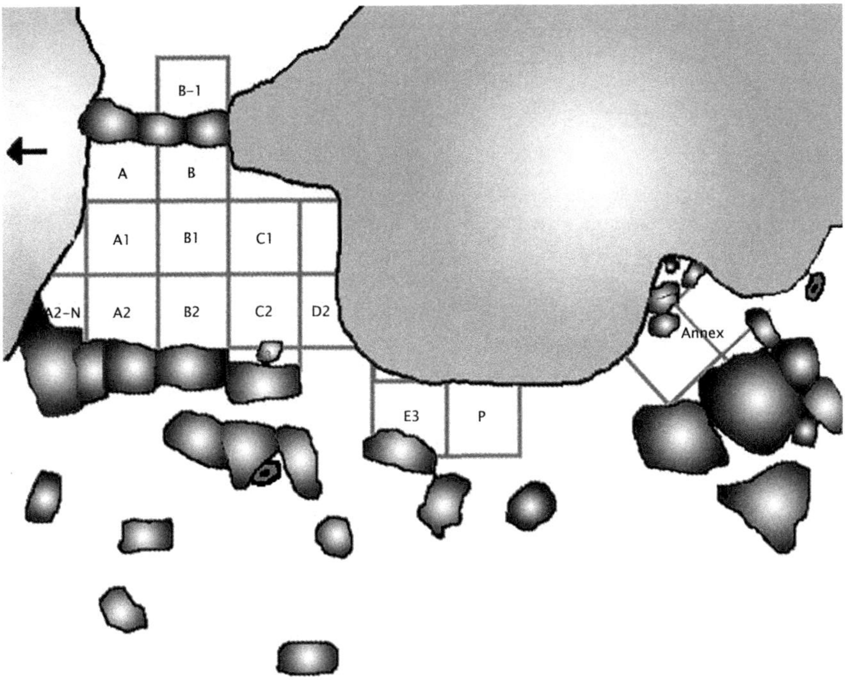

FIGURE 3.19 Site plan for Playa Sur 1, excavated 1995–96 (Drawing by Zarankin and Senatore 2007).

extent of discarded material suggested a medium to a long stay at the shelter. However, it was not possible to find any evidence of reoccupation.

Playa Sur 1 was one of the sites that showed a greater thickness of sediment layers (Figure 3.20). In most of the sites excavated, the depth from the surface to the sterile level is not more than 30 cm (5 cm in the case of Lima-Lima). However, on Playa Sur 1, this depth was 1 m (Zarankin and Senatore 2007):

- **Surface level**: It is made up of rocks detached from the cliff, a large amount of greenish algae and sand.
- **Level 1**: This level is composed of a more compact sandy sediment of lighter colours, mixed with algae. It has been interpreted by geologists as the result of decades of movement and natural accumulation of sediments (Claudio Parica pers. comm., 1995).
- **Level 2**: This level is typical of other sites with dark clayey sediment in which small fragments of bone appear, and is interpreted as the occupation floor. Most of the archaeological finds are in this level.
- **Level 3**: The last level, sterile, is a mixture of sand with pebbles and has no human action.

The 'Annex' was a semi-closed structure covering an area of 2 m^2. This building presented a large fire-pit contained by a circle of rocks

FIGURE 3.20 Excavations at Playa Sur 1, showing stone walls and whale vertebrae furniture (Photo by Andrés Zarankin and María Ximena Senatore 1997).

(Figure 3.21). Around the fire-pit, only two different artefacts were found: a shoe and a wooden stick possibly used to stir the contents of an iron pot. It is worth noting the striking contrast between the abundance and diversity of the findings in the shelter and the scarcity of artefacts in the annex. Possible interpretations of this and other small annexes associated with a number of other shelters include that they were working areas where some process involving fireplaces occurred (though these sites are too small and inconveniently located to be try-works); were cooking areas, either established while the larger shelter was being constructed over several days or a permanent kitchen area outside the main shelter (as cooking for a gang of maybe seven men using blubber fires would have been an extremely smoky and oily process).

Playa Sur 1 is another site that might relate to Captain John Davis' description of the distribution of sealing gangs along the Southern Beaches of the Byers Peninsula, presented earlier. If so, it might have been the middle of three camps spread along the coast under the control of Second Mate Phillips of *Huron* with possibly seven men. It would illustrate functional association with the sites at Elephant Point to the east and Sealers Hill/Vietor Point to the west, and would explain the large gaps between the campsites – the related gangs were spread out to maximise access to seal resources. This functional relationship would not be visible archaeologically unless material evidence was found in each of the sites that demonstrated their common origins, and in the absence of such material evidence, the documentary evidence is critical in suggesting this association.

- *Stackpole 1*

  62° 39′ 54.42″ S, 60° 57′ 10.26″ W

  Stackpole 1 is located on a wide moraine terrace, about 60 m from the sea, with no rock outcrops in its vicinity. It is characterised by two small structures, the largest about 3 m wide and 6 m long and the smallest about 2 m^2 (Figure 3.22). It has some rocks (possibly originating from the walls of the sealers shelter) and whale vertebrae (probably used as furniture) scattered on the surface of the land. It is likely that a large part of the site is sub-surface and some elements, such as whale ribs and vertebrae have been moved within the site or to areas adjacent to it by post-depositional factors such as animal disturbance, erosion and human intervention. The site was identified in 1995–96 by the Argentine project, and in 2012 the 'Landscapes in White' project excavated a small trench with no results.

  Due to its spatial and structural characteristics, very different from the majority of other sealer sites, two possibilities are suggested for its functioning. One possibility is that the surrounding area was a staking ground for the drying of seal skins, the collapsed stones being the remains of walls used

FIGURE 3.21 Playa Sur 1. Fireplace in 'Annex' structure (Photo by Andrés Zarankin and María Ximena Senatore 1996).

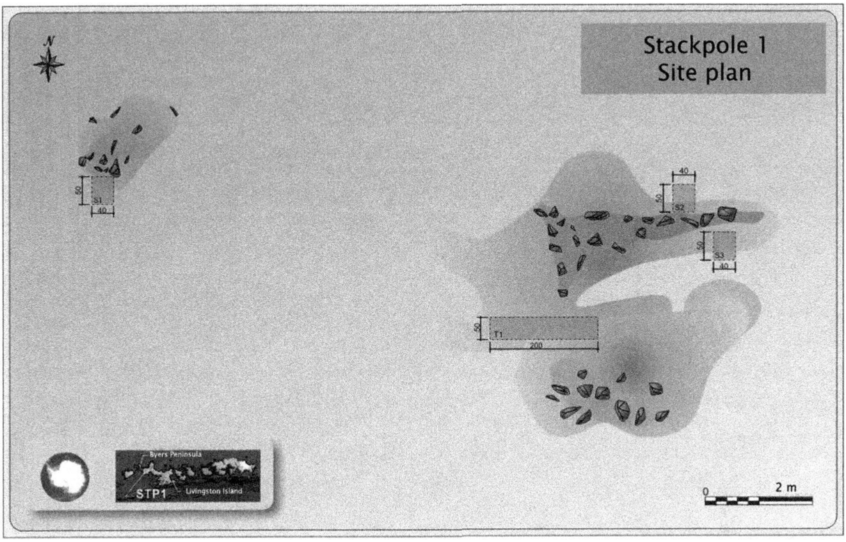

FIGURE 3.22 Site plan for Stackpole 1 (Drawing by LEACH 2012).

as temporary wind shelters while the sealers were working, or to protect the skins themselves. The wide and dry moraine terrace would have been a perfect location for such a function (Figure 3.23). The other possibility is that it was part of a grave or memorial for somebody's death. In any case, a major and more systematic excavation will be necessary to get a better understanding of the place in the future.

- *Stackpole 2*

  62° 39′ 55.86″ S, 60° 56′ 31.74″ W

  The site is on a wide moraine terrace about 70 m from the sea. It consists of a small rectangular structure about 2 m^2 that appears to be little disturbed, with other scattered elements (Figure 3.24). The site is marked by a slightly raised surface exposure of stones, one whale vertebra located among the stones and another three vertebrae on the surface in the vicinity of the site.

  This is another structure completely different from other sites in the region. Because of its slightly raised elevation and the superposition of several stones and bones (Figure 3.25), and because what might be a hole for a post, one possibility is that it was a grave site. Another alternative is that the stones and whale bones were used to support one side of an overturned boat providing an improvised shelter for sealers cruising the coast for seals (see Pearson 2018a). The site was not initially excavated to the extent necessary to address this possibility, and a subsequent survey in 2012 failed to locate any further evidence to clarify the site's function.

FIGURE 3.23 Stackpole 1 site overview and three-dimensional digital scanning activity (Photo by LEACH 2017).

- *Sealer 1*

  S 62° 40′ 26.9″, W 61° 06′ 47.6″

  The site is made up of two square-shaped structures sheltered by a large rocky outcrop (Figure 3.26), 80 m from the sea and 250 m from fresh water. In the larger structure, 0.65 m thick stacked stone walls enclose a 3.65 ×

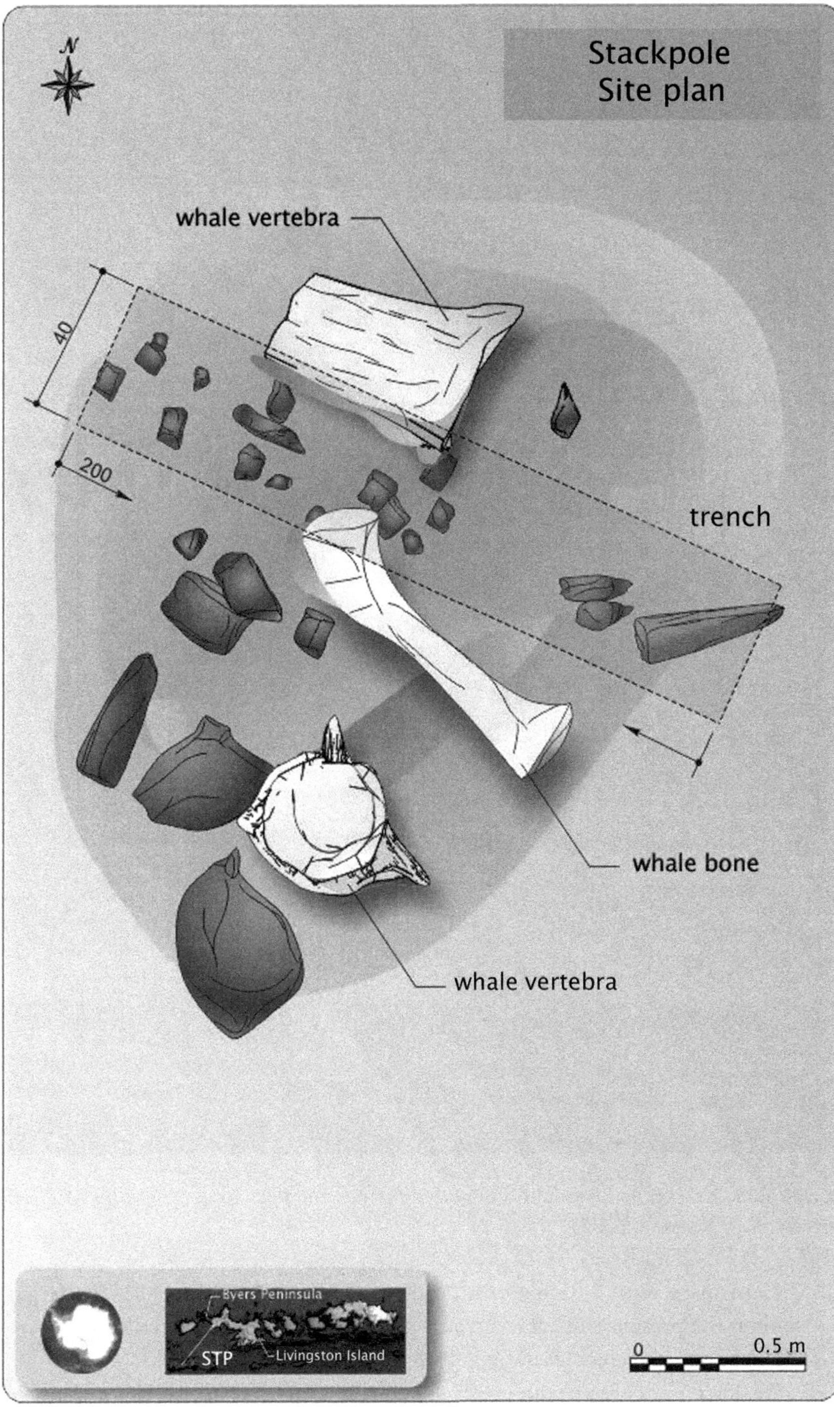

FIGURE 3.24 Site plan for Stackpole 2 (Drawing by LEACH 2012).

FIGURE 3.25 Stackpole 2 site, showing whale bones and vertebrae (Photo by Michael Pearson 2007).

2.45 m space with 8.95 m^2 of interior surface. Inside and outside the larger shelter, the presence of cetacean vertebrae is observed. The smaller structure is a 2.25 m^2 space enclosed by 1.40 × 1.60 m walls.

Sealer 1 is an area frequented by elephant seals, compromising site integrity. The interior of the larger enclosure is usually occupied by these animals,

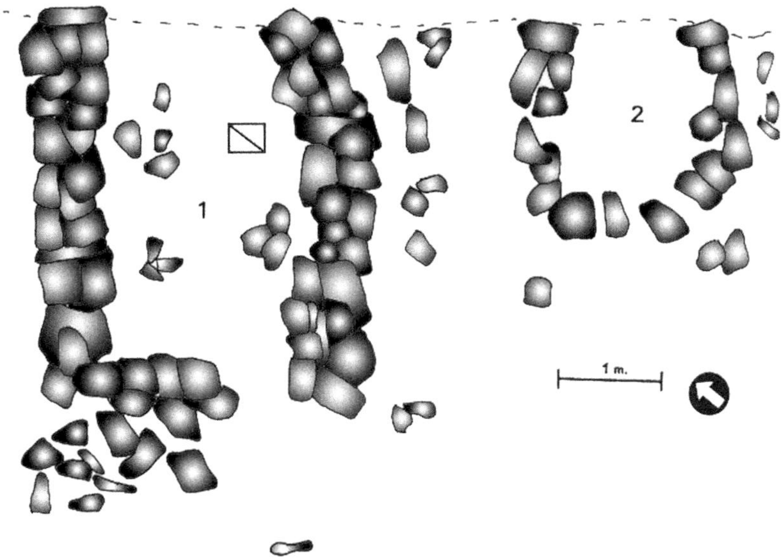

FIGURE 3.26 Sealer 1 site plan (Drawing by Zarankin and Senatore 2007).

which constantly erode the stratigraphy and have caused the walls to be completely collapsed.

The site was excavated in 2017. The collected materials included fragments of bone, leather, ceramics, metal and wood, which indicated the use of artefacts such as an axe, nails, clothes, pipes, improvised furniture and others. In total, about 70 artefact fragments were collected. All recovered materials received preventive conservation in situ, for transportation to the LEACH.

Four squares, called A1, A2, B1 and B2, with dimensions of 2 m^2, were laid out for the excavation of the main structure. In addition, testing was carried out in the second structure, the 'Annex', and two trenches to the east and west of it adjacent to its walls. The site seems to be typical of some others, with a structure devoted to accommodation and another one possibly oriented towards productive activities. There is no evidence that Sealer 1 is related functionally to other nearby sites.

- *Sealer 3*

  S 62° 40′ 23.1″, W 61° 06′ 12″

  The archaeological site consists of two square-shaped enclosures (Figure 3.27), against a large rock outcrop, 200 m away from the sea and 15 m from fresh water. In the larger structure, the walls of stacked stones, 1 m thick, enclose a space measuring 5.20 × 3.10 m, with 16.15 m^2 of internal surface. The smaller structure is a space measuring 13.20 m^2, enclosed by walls measuring 2.60 × 1.57 m.

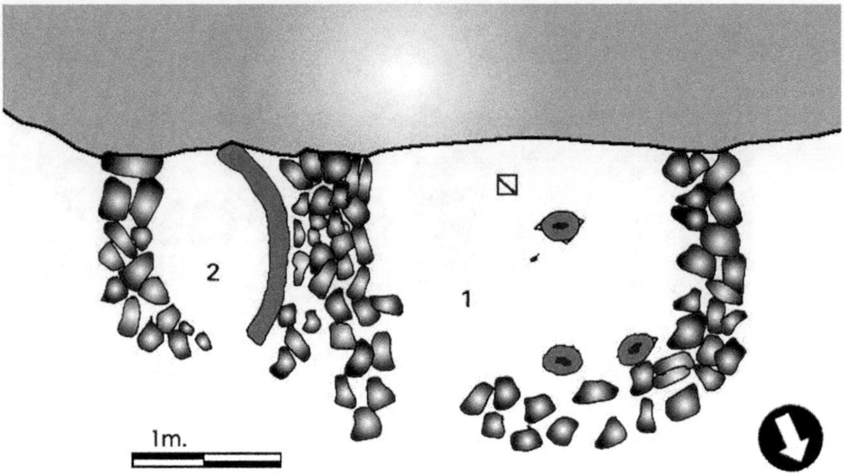

FIGURE 3.27 Site plan for Sealer 3 (Drawing by Zarankin and Senatore 2007).

In 2010, 75% of both structures were excavated (Figure 3.28). In the main shelter, approximately 20 cm below the surface, the occupation level was located, composed of a layer of dark clayey sediment, with a large amount of fatty organic content (Figure 3.29). Two fire-pits, next to which there were whale vertebrae (furniture of the sealers), appear to have provided warmth and cooking area, the space being interpreted as a living area and kitchen. The smaller structure had a roof built with a cetacean jaw, which acted as a beam. An accumulation of animal skins inside the structure leads to it being interpreted as a work and storage area for processed skins.

Among the artefacts found were shoes, clothing remains, canvas fragments interpreted as being part of the roof covering, smoking pipes, bullets, wooden stakes for skin-pegging in different stages of production and charred bones of various species. In one of the fireplaces, a fragment of hoop iron belonging to a barrel was found, which served as an improvised roasting grill on which small pans could be supported.

- *Sealer 4*

  S 62° 40′ 20.4″, W 61° 06′ 17.2″

  Sealer 4 site consists of two structures built against opposite sides of a rock outcrop (Figure 3.30). The larger structure has an interior surface of 19.25 m^2, closed at its ends by piled stone walls 1.60 m thick. The smaller structure is 3.5 × 3.5 m. Outside the larger structure, several whale vertebrae were observed. Materials collected on the surface included glass, bone and a smoking pipe (Zarankin and Senatore 2007: 78–79).

  The 2010s excavation evidenced a large combustion area (fire-pit) in the wall sector that borders the outcrop, configuring what at first would only be a work space (Figure 3.31). The recovered materials (piles, nails, gloves,

FIGURE 3.28 General view of Sealer 3 site (Photo by LEACH 2010).

bones, pipes, charcoal) appear to a depth of 20 cm, which also appears to be the floor of the settlement.

Due to the distance between Sealer 3 and Sealer 4 (approximately 80 m), their similar stratigraphic characteristics and shared types of materials, it is hypothesised that they were part of the same occupation from the 1820s, and that Sealer 3 functioned as a lodging and warehouse, while Sealer 4 was the place of work and processing of the hunted animals (Figure 3.32).

Sealer 3 and Sealer 4 are relevant as the first archaeological sites in Antarctica where geoarchaeological studies were conducted (Villagran and Schaefer 2011; Villagran, Schaefer and Ligouis 2013). Using multi-element chemical analyses, micromorphology and organic petrology, researchers were able to produce information not attainable through artefact or documentary analyses. Although local lithology (the characteristics of local rocks) and cryogenic (freezing) processes are dominant in the chemical and micromorphological records, respectively, important distinctions could be made, especially on the behaviour associated with the use of fire for cooking, processing and warmth and the use of local resources for survival. The results show the effects of human activities on the sediments and are expressed by higher phosphorus pentoxide (P_2O_5), calcium oxide (CaO) and total carbon (C) concentrations because of the use of seal bones, fat and imported wood as fuel for the hearths. Future use of geoarchaeological studies should make it possible to differentiate shelters with more intense occupation from single-activity sites and to show differences in the use of these structures,

FIGURE 3.29 Excavating materials and collecting sediment samples from Sealer 3 site (Photo by LEACH 2010).

for example distinguishing habitation shelters from working spaces for fat processing or storage.

The case studies discussed in this chapter illustrate the heterogeneity of archaeological sites in the South Shetlands and suggest ways in which the archaeological

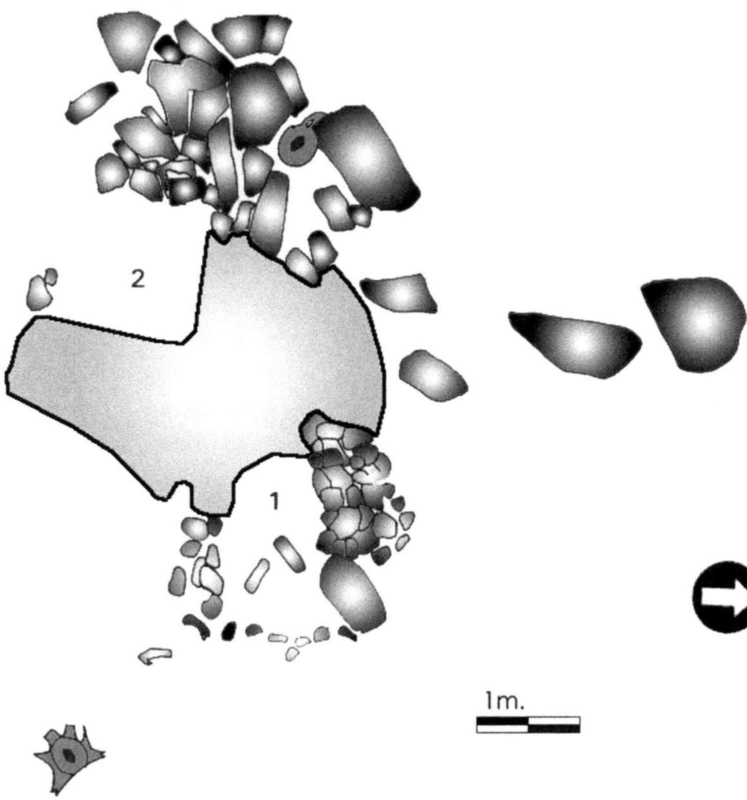

FIGURE 3.30 Site plan for Sealer 4 (Drawing by Zarankin and Senatore 2007).

evidence can cast light on the operational organisation of sealing and the lived experience of sealers. Some sites demonstrate considerable investment of time and energy, such as Lima-Lima and Punta Elefante 2, while others may be related to transient sealer visits, such as Stackpole 2. An implied relationship exists between complexity of sites and the duration of their occupancy. At some sites, the existence of 'annexes' or other shelters that appear closely related suggests separation of workspaces from living spaces. The nature of the use of these related sites remains conjectural – skin processing, oil boiling, domestic kitchens, storage of skins or equipment – but a pattern begins to emerge when the corpus of sites is considered. The importance of the documentary record in suggesting functional associations between separate sites is shown in the case of Punta Elefante 2 and Playa Sur 1, providing a cogent basis for assuming relationships that would be impossible to deduce from the archaeological record alone.

FIGURE 3.31 Sealer 4 site being excavated (Photo by LEACH 2010).

FIGURE 3.32 Image showing the relationship and distance between Sealer 3 and 4 sites (Photo by LEACH 2010).

We will discuss further in the next section ideas on distribution and analysis of materials, with a more complete analysis of the material evidence being provided in Chapter 4.

3.2.6 Spatial distribution of materials at sealers' camps

The most intensive study of sealing site artefacts has been the quantitative and qualitative analysis of the collection recovered by the 'Landscapes in White' project, led by Fernanda Soares, which was aimed at classifying objects and fragments and understanding their distribution at shelter sites. Special attention was paid to presences and absences, and the contextual association among artefacts, structures and other features. The study provided detailed information, giving a chance to discuss different aspects of sealers' life by focusing on the specifics of each site, and comparing to other locations (Soares 2020).

The analyses dealt with 4,139 fragments, which corresponded to the materials recovered in the 2010, 2011, 2012, 2014 and 2017 fieldwork seasons on Livingston Island. The remains collected in 2018 and 2019 at Lima-Lima were still being treated by specialists in conservation at the time of writing, so they were not included in the analysis (though they will be considered in the future). From the total surveyed so far (Table 3.1), the sites where larger numbers of fragments have been found included the following: Punta Varadero and Sealer 3, followed by Pencas 3, Punta Elefante 2 and Punta Elefante X3. Meanwhile, the vast majority of the findings were wood remains, followed by bones, textiles, leather, metal, skin, glass, kaolin, lithic (that is to say, stones that appear to have been used as tools) and ceramics. While these results describe the numbers of fragments recovered at the sites, it should be noted that studies on the minimal number of pieces (denoting the smallest number of artefacts represented in the collection) were also conducted with an aim not to overestimate the presence and relevance of materials prone to high breakage (wood and bone, for example, falls into this category).

From the quantification of the materials and the analysis of their distribution using plans drawn in the field, it was possible to create maps of concentration of the different remains using the SYSTAT program. A series of patterns are represented in several sites, and they provide interesting information to interpret some practices. The distribution of wood fragments, sometimes at the periphery of a site, might indicate the chopping of wood for use in the fireplace. Other materials, such as textile, metal, leather and bone fragments tended to be found on more central locations. Materials were generally found in proximity to fireplaces and whale vertebrae. Fireplaces were the source of heating and cooking, while whale vertebrae were the 'kitchen table' and seats around which the life in the shelter focused: a place to relax, prepare food at the adjacent fireplace, repair clothing, smoke and chat. Further interpretation of life within the shelters is given in the next chapter.

TABLE 3.1 Number of fragments by sites and types of materials

Number of fragments

Site	*Tableware*	*Kaolin*	*Fabric*	*Leather*	*Bone*	*Skin*	*Wood*	*Metal*	*Glass*	*Lithic*	*Others*	*Total*
Sealer 3 (large structure)	0	1	215	41	212	5	356	63	0	23	90	1006
Sealer 3 (small structure)	0	0	0	1	2	12	37	1	0	0	1	54
Sealer 4 (large structure)	0	4	77	2	32	9	267	12	0	7	16	426
Sealer 4 (small structure)	0	0	0	0	0	0	10	0	0	0	0	10
Pencas 3 (large structure)	0	54	65	43	68	19	200	115	77	24	7	672
Pencas 3 (smaller structure)	0	7	0	0	2	0	1	4	1	0	1	16
Punta Varadero (large structure)	1	3	294	229	242	30	24	69	8	2	18	920
Punta Varadero (small structure)	0	0	0	1	13	24	0	0	2	0	0	40
Punta X 1	0	2	59	1	119	75	196	12	0	3	93	560
Punta X 2	0	22	30	10	198	1	141	24	8	0	1	435
Punta X 3	0	0	0	0	0	0	2	6	0	0	2	10
Punta Elefante 2	0	7	8	9	399	0	87	105	37	2	1	655
Punta Elefante X1	0	0	110	30	5	19	182	53	3	17	0	419
Sealer 1	0	1	10	1	110	16	65	77	0	0	24	304
Total	1	93	740	328	888	175	1224	306	96	59	229	4139

3.2.7 Sealer's experiences of landscapes and life at shelters

The 'Landscapes in White' project has manifested a particular interest in discussing how sealers incorporated the South Shetlands territory. Traditional approaches have largely relied on cartographic perspectives highlighting the role of reason and the sense of vision in human relationships with the region (Salerno et al. 2010; Salerno and Zarankin 2014; Zarankin and Salerno 2018b). Without diminishing the relevance of these contributions, the project has also considered the importance of sealers' bodily experience as a means to incorporate the region in a practical way. Here the word 'incorporation' refers to the possibility of gaining familiarity with the archipelago, making it flesh and part of oneself. Influenced by phenomenology and the theory of embodiment (Csordas 1990; Merleau-Ponty 1993), the project has considered the role of culturally informed sensorial, mnemonic and affective bodily regimes to shape human encounters with the territory (Salerno 2011; Salerno 2014; Cruz 2019).

Here, it is worth noting that sensory experience is multiple and synesthetic, being culturally embedded with meaning and values and having an impact on the construction of affective impressions and memories (Howes 2006; Hamilakis 2014). With an aim to approach sealers' experiences in the past, archaeologists sometimes resort to their own sentient bodies as a relevant tool to pose questions. However, researchers' own experiences cannot be projected into the past as experience is always culturally situated. Having a deep understanding of sealers and seamen's lives then becomes fundamental to guide the analysis. Historical accounts as well as a contextual analysis of the material world of these people help reconstruct the sensorial regimes that could have shaped their perceptions (Hamilakis 2013). The approach does not deal with individual experiences but with the inter-subjective domain that frames people's cultural ways of engaging with the environment.

When people experience the surrounding world, some things may come to the fore while some others remain in the background (Merleau-Ponty 1977). The things that actually come to the fore represent 'sensitive points' or 'agent images' that actively call for their attention, demanding to be approached in specific ways (sensu Besse 2006; Yates 2007). Sealers approached the territory both as hunters and as seamen, so the geographical features of the coasts and the presence of animals were key to their perception of landscape. Weather – including cold, high winds, etc. – also had an impact on them, as well as on other people visiting the region. However, these factors should be equally understood within the context of sealers' material conditions of existence. Sealers did not only know the South Shetlands from a distant gaze but also from the efforts required to circumnavigate the islands and disembark and walk on pebble beaches that destroyed their shoes and exposed them to cold and humidity. Sealers did not just relate to animals in an objective way, but hunting and processing compromised their bodies and senses in specific ways, from the bloodsheds to the smoking try-pots.

Researchers have the possibility to identify some of the features of the landscape that could have shaped sealers' experiences on the areas they actually operated. Moreover, the material remains of shelters and the associated artefacts recovered there might provide interesting clues on sealers' experiences of inhabiting these places. The analysis of the dimensions of the structures, and the number of people that could have possibly inhabited them, has led archaeologists to think of sealers' shelters as reduced and crowded spaces that often demanded people to be sat or lying rather than standing, leaving small room for personal space. The lack of segmentations within the space of the structures created an appropriate atmosphere for sharing sensory experience. Comparisons with the material conditions of vessels rooms have provided an interesting counterpoint to discuss sealers' existence in different scenarios.

3.2.8 Public archaeology approaches to sealers' life

Academic archaeology has traditionally had enormous problems in establishing communication with the public. This situation is related to the technical language researchers use, their possible lack of imagination to tell appealing stories and an authoritarian perspective of considering themselves the owners of truth about the past. Public archaeology has for some time been reflecting on how to resolve this 'top-down' approach and democratise the situation. Some of the questions public archaeology deal with include the following: How can researchers make their narratives accessible rather than technical and scientific? How can they get around the lack of prior knowledge? How can they make the concepts and approaches more easily comprehensible? How can they give the public the opportunity to consider the issues without 'specialists' interpreting them?

In this context, mediation becomes a central point to provide alternative forms of scientific divulgation and knowledge construction (Soares and Martire 2021; Zarankin, et al. 2021). One of the attempts to achieve this goal is the use of digital technologies as potential tools to transcend the crevasses that separate academic disciplines and community (Richardson 2009). Resort to several immersive and sensorial strategies, such as the recreation of environments or installations within enclosed spaces and an element of spectator participation is also being tested.

In the case of the 'Landscape in White' project, these new approaches to public archaeology have been developed to try to make the Antarctic sealing sites more accessible and meaningful for audiences. These sites are extremely remote, almost never visited by anyone other than researchers, are quite enigmatic in their physical remains, and do not fit easily into the historical narratives of any particular country. It is not surprising that the sealing sites are little known, even though the general public is fascinated by Antarctica more generally. The 'Landscapes in White' project has made efforts to make archaeological findings widely available to the public and to stimulate the entanglement between the people of the present and 'the things of the past' (sensu Bezerra 2013). It draws upon the widespread general interest in Antarctica and uses the potential offered for presenting that unique environment to underpin the story of the sealers and their experience of the frigid south.

One demand on any public archaeology program is to make available well-researched and accurate information. In a digital age, public archaeology programs can parallel the role of museums, by providing access to sites, their archaeological study and the associated artefacts in digital and augmented reality forms. The 'Landscapes in White' program is a good example of this in action, with an online database of sites and artefacts backed up with 3D imagery (Figures 3.33 and 3.34). Much of this content can be found online (http://www.leach.ufmg.br; https://sketchfab.com/leach.ufmg).

Another approach is to engage the public audience in new ways that break out of the traditional pedagogic models. The 'Landscapes in White' program has developed an approach that engages the sensory involvement of the audience in a simulated Antarctic environment (Figures 3.35–3.37). In exploring ways of eliciting a sensory response to the Antarctic experience, the program is seeking to utilise new types of responses and connections to the territory through techniques such as sensorial exhibitions and domes, theatre performances, digital media, 3D models, etc.; and seeking to strengthen the communication channels between archaeologists and general public, implementing activities that are exciting, meaningful, symmetrical and less directed (Shanks 2007; Soares and Mota 2017; Soares et al. 2018).

In 2017–18, a sensorial exhibition was prepared by Mediantar, Mycoantar and the 'Landscapes in White' research projects, and presented at the UFMG Museum (Belo Horizonte, Brazil). The exhibition made it possible to show the work of the three projects and to collaborate with the people from the Museum to find alternatives to engage with the public through sensory experience. In addition to texts, images, videos and sounds, sensory capsules were used to simulate the experience of walking on the snow, wind and darkness, etc. A 360°

FIGURE 3.33 3D scanning of Lima-Lima Cave using laser Scan 3D Leica P20 (Photo by LEACH 2018).

FIGURE 3.34 Virtual model of Sealer 1 site created from 3D scanning (Photo by LEACH 2018).

FIGURE 3.35 Antarctic exhibition at the Federal University of Minas Gerais, Brazil, where some of the results of the 'Landscapes in White' project were presented to the public (Photo by LEACH 2018).

video about the continent was projected at the UFMG Museum planetarium, and various educational activities and workshops for children were prepared by the Museum's pedagogical team. More than 50,000 people visited the exhibition (Zarankin and Soares 2021).

The archaeology dome was an acclimatised inflatable capsule that held, inside, a life-size replica of the nineteenth-century sealers' site Punta Elefante 2. This work was carried out in partnership with the team from the Pedagogical Centre and Ponto Museum at UFMG. The dome was entirely designed to resemble the experience we researchers had at the archaeological site. Air-conditioners were installed to keep the temperature low, so the visitor could experience the cold of the continent; speakers were used to reproduce the sounds of marine animals and high southern winds; a lighting system was placed to recreate the luminosity and brilliance of Antarctica; 180° video projectors were played to provide a more immersive view of the Antarctic landscapes; etc. These features intended to encourage a more synesthetic experience of the South Shetlands landscape (Soares et al. 2018).

A theatrical performance represented a fictional conversation between a nineteenth-century sealer and a student of archaeology. The sealer's costume was prepared by a fashion designer, inspired by archaeological and historical findings. The script involved a series of questions researchers would like to make sealers if they had the chance to meet them, while material objects were the conductors of the conversation. In recent years the 'Landscapes in White' project also produced a sticker album for children, a comic book and tested the use of augmented reality, 3D-printed replicas of artefacts and 3D virtual models, among others, as potential tools for dialogue with the public (Soares and Mota 2017; Soares and Zarankin 2020; Soares and Martire 2021).

FIGURE 3.36 'Sensory dome' where visitors could have a simulated experience of sealers' life at Antarctic shelters (Photo by LEACH 2017).

FIGURE 3.37 Videogames with educational purposes about Antarctic archaeology (Photo by LEACH-ARISE 2021).

The product, called 'Antarctic Adventure – Archaeology in the frozen continent' (Figure 3.37), developed by Alex Martire has been developed in partnership with the Interactive Archaeology and Electronic Simulations Group (ARISE-MAE/USP). It fits in the area of 'archaeogaming' (Reinhard 2018), since it unites the two main trends of this perspective: on the one hand, it is a game that portrays the life of nineteenth-century sealers in Antarctica, and on the other hand, it represents contemporary excavations carried out there. The videogame offers the player the possibility of fulfilling missions in the past and present, in order to establish correlations between the daily life of sealers and the artefacts that archaeologists find in the South Shetlands today.

3.3 Final words

The study of the spaces occupied by sealers and of the material culture with which that occupation was achieved offers the opportunity of interpreting the connections established between people and Antarctica as the expansion of the modern world incorporated that new and challenging territory. The archaeological evidence is a necessary tool in the analysis of the sealer's adaptation to the new environment in conditions of forced self-sufficiency. It enables a reinterpretation of the history of the sealing period that de-emphasises the role of the company owners and ship's captains, dominant in the historical literature, and gives life to the story of ordinary sealers who did the work of exploiting marine mammals for profit, work undertaken under extreme conditions and material hardship. It shows how the combination of texts, maps, sites and objects can be understood as equally eloquent fragments of past experiences. While the contemporary narratives articulate the why and when of the sealing venture and provide the framework for understanding the context within which it occurred, the archaeological record will lead to the birth of new and more inclusive histories of the first human interactions with Antarctica. The next chapter expands on the way the practices and experiences of sealers can be understood better through the combination of disparate historical and archaeological evidence.

4

APPROACHING SEALERS' PRACTICES AND EXPERIENCES

Since the 1980s, research in archaeology has presented the opportunity to learn more about sealers' lives on the South Shetland Islands as well as at some other sealing grounds in the Southern Oceans. While ashore, the men devoted time and efforts to working, subsistence and leisure time activities. Historical and archaeological evidence provides information on the material culture of both branches of the sealing trade, exploiting fur seals and elephant seals, and enables the particularities of hunting and processing in the region to be discussed. When sealers remained at a given location for more than a few hours, and work was done, they had to find ways to subsist in a harsh environment. Material findings prove relevant in shedding light on a wide number of related issues, including sheltering, sleeping, dressing, eating, drinking, smoking, etc., not always thoroughly contemplated by documentary sources.

In the previous chapter, we presented the archaeological background of the nineteenth-century sealers' sites in the region, considering the research activities carried out over time, the analysis of the diversity of sites and a general description of the material culture identified at these places. In this section of the book, we will particularly draw on this evidence to discuss a variety of daily practices and experiences that help us to build a better understanding of the history of these people.

4.1 Working practices

The processes of hunting and preparing fur seal pelts and elephant seal oil are outlined in Chapter 3. During the nineteenth century, strategies for sealing on the South Shetland Islands, and indeed at all the sealing grounds in the Southern Oceans, evolved. Transformations included the shifting of focus from fur seals to elephant seals as the former became scarce, and the changing duration of stays

DOI: 10.4324/9780429201257-5

ashore found necessary to obtain the product. These changes were influenced by the growing geographical knowledge of the region, the fluctuations in the relative abundance of animal resources, technological developments, market changes to other furs or materials and to petroleum oils and other factors (Salerno and Cruz 2019). Throughout the period, hunting and processing fur seals and elephant seals depended on the ability to land men and equipment on beaches from the sea, access by land being interrupted by glaciers or demanding unacceptably long traverses over rocky and boggy terrain. Boats and tenders were key to carrying out these operations.

The pattern of ship operations at sealing grounds has been described previously – large ships brought expeditions from the home port, lighter tenders called 'shallops' either accompanied them or were built from prefabricated timbers at a convenient island and were used as work-horses servicing the dangerous-to-approach beaches. Light rowing boats, often whaleboats, carried the men ashore and resupplied them, retrieved skins and oil, and were used for exploring new beaches. This diminishing capacity in the vessels making up the supply chain of ship, shallop and boat explains why the equipment and building materials landed at the sealing sites were very limited in size and amount. Try-pots were probably the largest item able to be carried ashore, and in most cases, larger casks filled with oil had to be towed out to the shallop as they were too heavy and dangerous to carry in the boats.

The British and the American markets were always a central component of the fur sealing trade. China, through Canton, was a key market from the 1780s, but the British ceased trading skins there in 1803, and the Americans by the early 1830s. The British and American sealing fleets were about equal in number until the late 1820s, but British sealing virtually disappeared by about 1840, leaving the United States as the dominant supplier through to the 1880s. The elephant seal oil trade was more stable, being in step with the pricing for whale oil, until its decline at the end of the century, largely due to its replacement by mineral oil and gas (Richards 2003: 2; Basberg and Headland 2008: 6–8, 18). London was a dominant seal trading market by the late eighteenth century, in large part because of the discovery in the 1790s of the system for removing the guard hairs and allowing the underfur to be gathered for felting, a process previously known only to the Chinese. However, the importation of fur seal skins to London dropped dramatically at the end of the 1820s (Busch 1985; Smith 2002; Basberg et al. 2018; Burton 2018).

The preference on the sealing markets for dried or salted furs, and the influence that has on the processes carried out at the sealing grounds, is as yet not well researched, and the evidence is ambiguous. As mentioned above, the Chinese market is said to have preferred dried furs, but the major historian of the later nineteenth-century American sealing, A.H. Clark, claimed that US sealers stopped drying skins by 1815, some 20 years before it stopped trading fur seal skins to Canton (Clark 1887: 430). To complicate the issue further, sales figures indicate that dried fur seal skins still held a higher price than salted skins in New York at least in 1823 (Basberg and Headland 2008: 22 table 2). In the South Shetlands, both salting and drying of skins were common in the 1820s. It is clear, however, that by at least the 1840s salting was the dominant process for the greatly diminished American sealing trade, and voyages via the Cape Verde Islands to collect salt and labour were the norm. The British had left the trade by that date. As seen in Chapter 2, the presence of Cape Verde sealers also had an influence on the archaeological footprint at elephant sealing grounds such as Heard Island.

The tools used in sealing were well described in the literature, and a few examples have been found in archaeological excavations or as surface finds (rare as they were generally taken back to the vessels so they could be used somewhere else) (Figure 4.1). Clubs, used for both fur seal and elephant seal killing, seem to have varied in form. Ames recounts that: 'each man was armed with a formidable club, nearly six feet long, such a piece of timber as a farmer would cut for a "sled stake", a swinging blow from which would have overturned the ox Columbus' (1830: 132). Other accounts describe sealing clubs varying in length, from 18 inches (46 cm) to three feet (91 cm) up to Ames six feet (1.8 m) long (Townsend

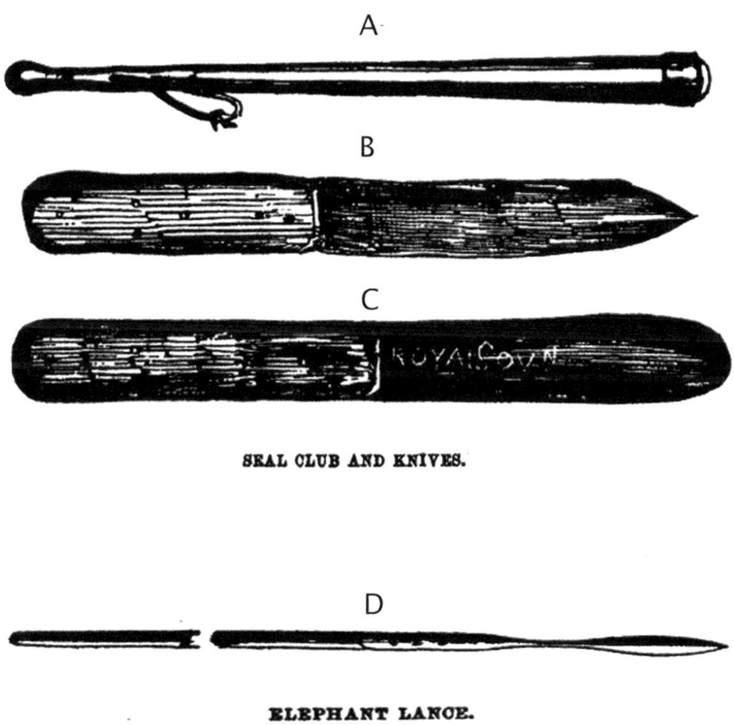

FIGURE 4.1 Sealing equipment, as used on Kerguelen Island in the 1820s: (a) Sealing club with a metal band on the head and wrist cord; (b–c) Pointed and round-tip knives; (d) Elephant seal lance (Source: Clarke 1850: 26–27).

1797 quoted in Delano 1817: 306; Clark 1887: 435; Starke 1986: 27). The end of what may be a club, with an iron band around the tip was excavated at the Sealer 1 site on Livingston Island, South Shetlands in 2017, and several probable clubs were surface finds on the nearby beaches (also see Zarankin and Senatore 2007: fig 151). These complete clubs were about three feet long, matching Townsend and Delano's descriptions.

Even though all of the wooden clubs show a distal portion thicker than the handle, each one presents slight differences in shape – ranging from cylindrical to spatula-type forms (Figure 4.2). These particular features could have been a result of the hand-made manufacturing of the pieces and/or some of the post-depositional processes affecting the remains. The metal rings fixed to the wooden clubs by means of iron nails could have been used to increase the effectiveness of sealers' blows (Nolasco 2018b). The drawing depicting a wooden club in Clarke's *Narrative of the Wreck of the Favorite* (1850: 26) reveals the presence of these metal pieces, as well as a strap or string on the handle, presumably to go around the wrist so the club could be let swing free when other tools such as knives were being used. It is of note that one of the clubs found on the Southern Beach of Byers Peninsula has red stains on its head-end. Further studies will be required to determine if the substance involved is blood stain or is from some other cause.

In all, the 'Landscape in White' project has recorded a minimum of seven wooden clubs since 2009 (pieces 'A', 'B', 'C', 'P', 'Q', 'R' and 'T') (Table 4.1). These have different lengths (22 cm; 19 cm; 18.6 cm; 56.1 cm; 100 cm; 89.5 cm and 66.3 cm) and varied widths (2.5 cm; 2.5 cm; 2.7 cm; 8.8 cm; 10.8 cm; 7.3 cm; 4.2 cm). Pieces 'A', 'B' and 'C' are from the Punta Elefante X1 site, parts 'P', 'Q' and 'R' are not from a specific site but were found on the Southern Beach of Byers, and the 'T' piece is from the site Punta X3.

Knives were the other main tool of the sealer (Figure 4.3). All seamen carried sheath knives, which were essential for jobs aboard ship (such as repairing or freeing rigging) and could be carried ashore for a wide variety of tasks. Nathaniel Ames (1830: 131–132) recounts that '… each man was armed with a … case containing three very large knives, one with a blade more than a foot long and others somewhat smaller, made of excellent steel'. On Kerguelen, the knives were carried in a similar case

> formed of two pieces of wood hollowed into three chambers, so as to contain the steel, flencing knife and ripping knife, and the two sides pegged or strapped together: this case was attached to our left side by a strap around the waist.
>
> *(Clarke 1850: 129; drawing p.26)*

A knife case very similar to that described is shown in a painting by Augustus Earle of skinning a seal on Tristan da Cunha in 1824 ('Flinching a yearling, a young sea elephant, Tristan de Acunha', National Library of Australia, https://

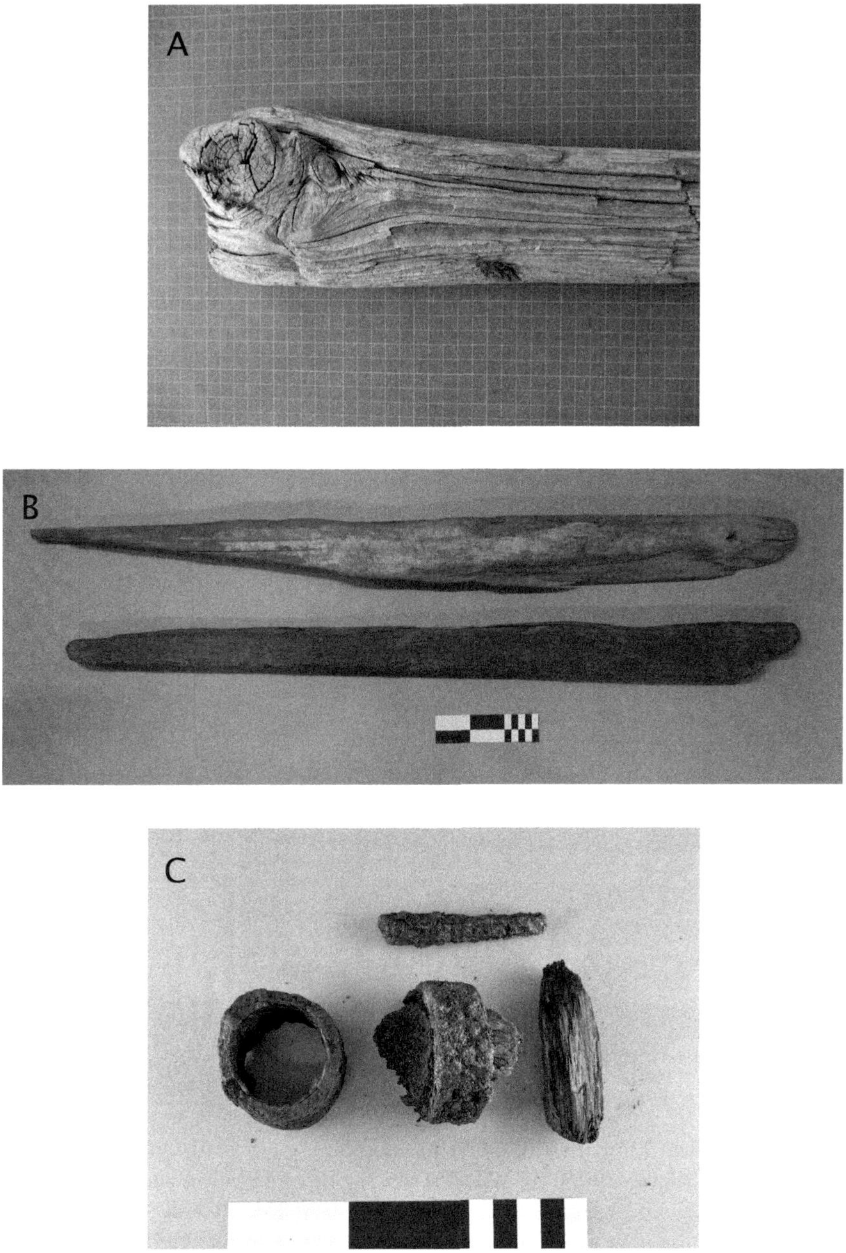

FIGURE 4.2 Archaeological remains of seal clubs found on Byers Peninsula, South Shetland Islands: (a) Head of a seal club recovered in the 1990s (Photo by Melisa A. Salerno 2021); (b) Probable seal clubs collected on the Southern Beach; (c) Metal rings associated with the head of a wooden club found at the Sealer 1 site (Photo by LEACH 2022).

TABLE 4.1 Sealing clubs from the LEACH Antarctic collection

Sealing clubs					
Id.	*Max. length*	*Max. width*	*Diameter*	*Shape*	*Site*
A	22,0 cm	2,5 cm	2,5 cm	Cylindrical	Punta Elefante X1
B	19,0 cm	2,5 cm	2,5 cm	Cylindrical	Punta Elefante X1
C	18,6 cm	2,7 cm	3,0 cm	Cylindrical	Punta Elefante X1
P	56,1 cm	8,8 cm	4,5 cm	Triangular prism	Southern Beach
Q	100 cm	10,8 cm	6,2 cm	Triangular prism	Southern Beach
R	89,5 cm	7,3 cm	2,7 cm	Rowing	Southern Beach
T	66,3 cm	6,2 cm	4,2 cm	Parallelepiped	Punta X 3

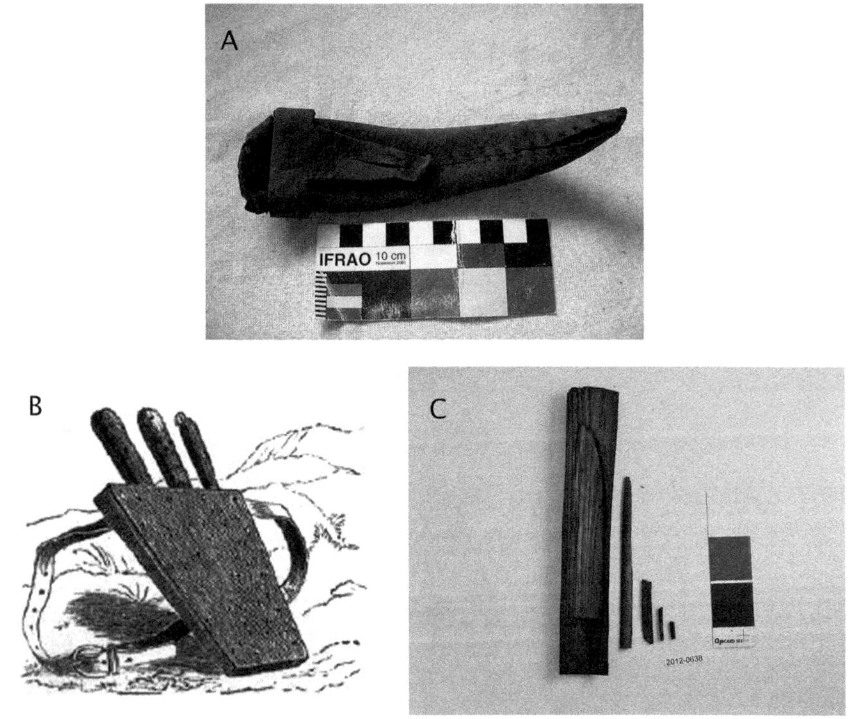

FIGURE 4.3 Sealers' knife sheaths. (a) Archaeological leather sheath (Photo by Melisa A. Salerno 2007); (b) Wooden case for carrying knives and honing steel, as used on Kerguelen Island in the 1820s (Source: Clarke 1850: 145); (c) Archaeological case and pegging stakes (Photo by LEACH 2018).

nla.gov.au/nla.obj-134494616/view), and an example of one was excavated at Lima-Lima Cave on Byers Peninsula in 1958, having two pieces carved with indentations to hold a large knife and a sharpening steel, and two indentations around the outside for strapping the two halves together (Lewis-Smith and Simpson 1987: 63). Another half piece of a wooden sheath for one knife was excavated in 2012 at site Punta X3 on the Southern Beach of Byers Peninsula, and another at Punta Elefante 2 (Soares 2020). Despite some differences in the number of chambers, the wooden cases described in the historical literature and found in archaeological contexts are essentially identical. In the 1990s, the Argentine project recovered a leather knife sheath at the Cerro Negro site on Byers Peninsula.

Also found in the excavation of Sealer 1 was an axe with an iron blade and a wooden handle. Pieces of skin were still adhered to the blade (Nolasco 2018b). This artefact could have been used for killing and/or processing marine mammals, which were not necessarily valued for their fur (such as elephant seals), or for butchering a carcass for domestic consumption. What appears to be a pike-style boat hook was recovered in the 1990s (Figure 4.4), which may have been used in the style of an Arctic sealing 'hakapik' to recover seals from water around rocks or to drag blubber across the beach.

Processing activities for fur seal pelts have left a specific archaeological evidence. Wooden pegs for skin drying have been found in several archaeological sites on the South Shetland Islands (e.g., Pearson and Stehberg 2006: 344; Zarankin and Senatore 2007: fig. 152). The relative abundance of these findings matches historical references that point out that about 10 to 14 pegs were necessary for stretching a single skin. As many of these objects were found inside sealers' shelters, it seems likely that they were manufactured there while sealers were not hunting. The men could have used wood brought with them from the sealing

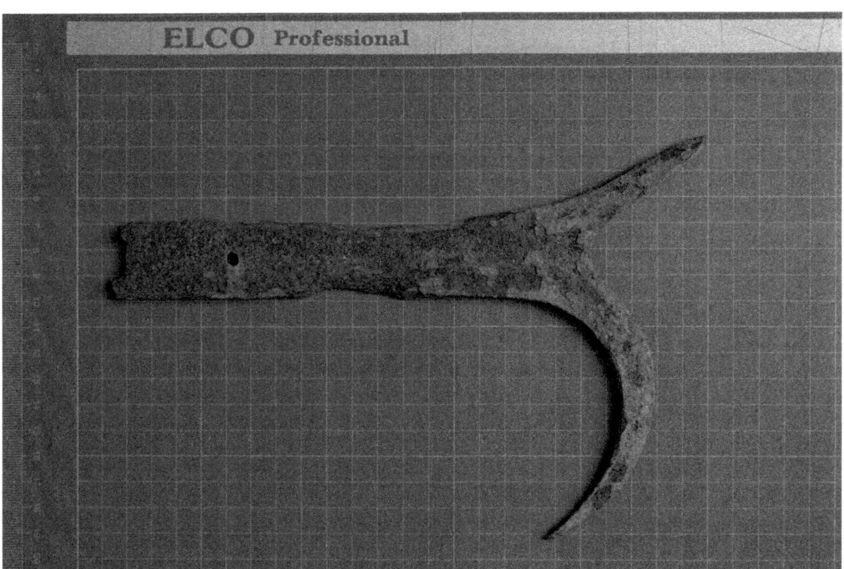

FIGURE 4.4 Iron boat hook recovered at Punta Varadero, possibly used to recover seals from water around rocks (as in the use of the 'hakapik' in Arctic sealing) or to drag carcasses or blubber across the beach (Photo by Melisa A. Salerno 2021).

ship to prepare these articles or even wood coming from shipwrecks or wrecked boats. Considering the cut and chipping marks on the surface of some items, knives or other sharp instruments appear to have been used to produce them.

Wooden pegs found in archaeological contexts are pointed and generally between 10 and 20 cm long, showing slight differences in shape (including cylindrical, prismatic and parallelepiped pieces) (Figure 4.5). Phelps records the crew making the foot-long (30 cm) pegs on the voyage south (Phelps 1871: 29–30). Many of the artefacts are broken, and some of them present a dark shade on their tip, probably as a result of having been buried in the ground to stretch the skins (Nolasco 2018b). However, wooden pegs are seldom found on the beaches, due to two centuries of soil movement, animal traffic and natural decay. Likely pegging grounds have been identified in the South Shetlands, a possible one on Spark Point on Greenwich Island had a square stone structure 1.5 x 1.5 m, that may have been used for stacking dried skins (Pearson field notebook 2007, see Appendix). Stacked skins found archaeologically at the Playa Sur 1 and Sealer 1 sites on Livingston Island may have been stored for future transport to the ship but for some reason left at the site (Zarankin and Senatore 2007: 111).

In the 1990s, the members of the Argentine project found a scraping tool made of glass during a surface collection at the Cutler 1 site on the North Beach of Byers Peninsula (Figure 4.6). It is a piece from a broken bottle, flaked to produce a robust sharp edge on one of its sides. This technique could have been similar to the one employed by indigenous groups from the southern end of South

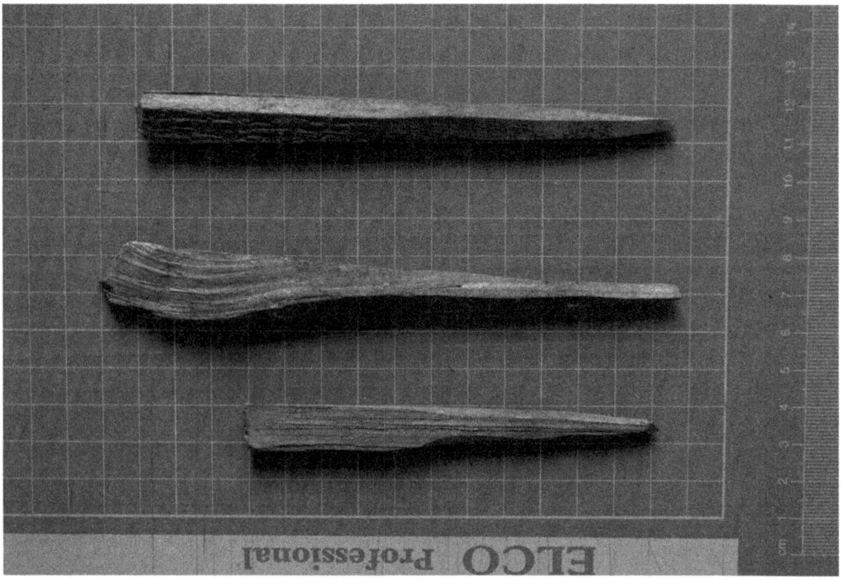

FIGURE 4.5 Wooden pegs, probably cut to peg-out seal skins for drying in the sun. Many of these have been found in sealing sites (Photo by Melisa A. Salerno 2021).

America, as confirmed by the Argentine lithic specialist Carlos Aschero (pers. comm. 1998). The tool could have had several uses, including the preparation of seal skins or clothing. Lithic tools have been found at some other sealing sites on the South Shetlands, including Cora B and Cora D on Desolation Island. The origin of these artefacts has been discussed in the literature, considering whether they were gunflints or indigenous artefacts (Stehberg and Blanco 2002), concluding that they were in fact of indigenous origin based on morphological and technological grounds, the patterns of use and the presence of organic remains on the surface (in particular, animal fat). In one case, some examples were shown to have been artefacts from central Chile maliciously inserted into coastal marine samples from Admiralty Bay in the 1970s by persons unknown (Stehberg 1983b). As discussed in Chapter 3, the others, however, appear to relate to indigenous workers accompanying the sealers in the 1820s, or in any case to the knowledge of the techniques for producing this kind of tools.

Archaeologists have found piles of rocks next to some sealers' sites that do not appear to be associated with the construction of shelters. Although the use of these piles is still to be researched fully, it has been speculated that the rocks could have served the seal skin processing. Historical references for the Cape Horn region suggest that sealers built up an area with rocks in order to stack skins, let them sweat and allow the water drain under them (Athearn in Clark 1887: 434). There are also references to 'pressing' such stacks of skins for a day or two to squeeze out water before salting (Clarke 1887: 430). Layers of buried seal skins have been sometimes found in small structures possibly used as storage

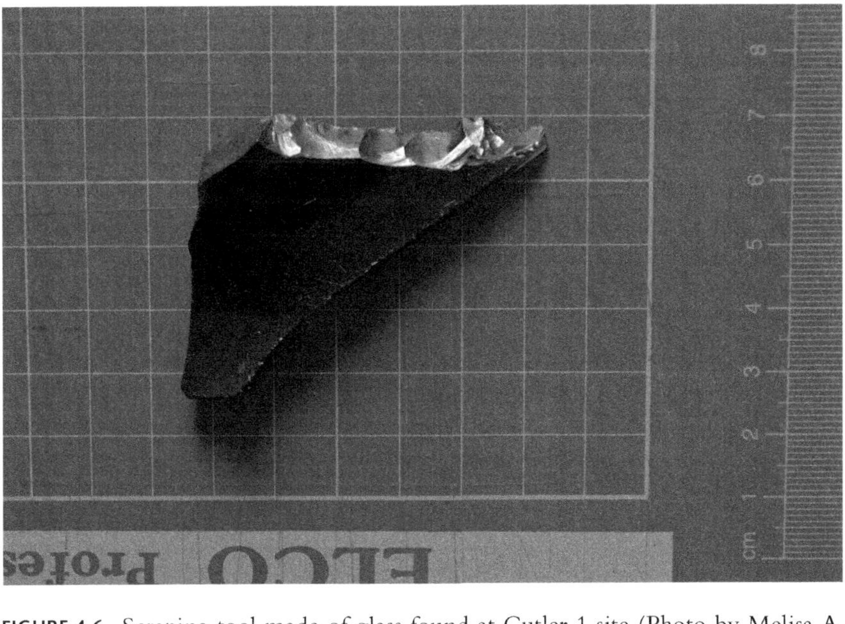

FIGURE 4.6 Scraping tool made of glass found at Cutler 1 site (Photo by Melisa A. Salerno 2021).

rooms (Zarankin and Senatore 2007: 111; Villagran and Schaefer 2011; Villagran et al. 2013). The exact reason for leaving seal skins behind remains unknown, as time and effort were invested to take them. On the one hand, sealers could have faced difficulties in taking the skins on board, such as bad weather at the end of a season, leading to the decision to abandon all or part of the collected skins. On the other hand, sealers could have left the skins in the storage rooms with a view to come back later but never have had a chance to do so. A similar case of the forced abandonment of sealing produce is referred to in Chapter 2, when 35 barrels of elephant seal oil were left ashore at Oil Barrel Point on Heard Island because the US naval ship that rescued the sealers, whose own ship had been lost, would not take them aboard.

Elephant seals, being much larger animals than fur seals, were harder and more dangerous to take. As presented in Chapter 3, the men approached the animals while they were lying on the beaches and knocked them down (as sealers did in the case of fur seals) or made them sit up by striking them with a wooden club, depending on the size of the seal. After stunning small animals or making larger ones rear up to expose their chest or open their mouth, the hunter either lanced them or shot them. Nathaniel Ames in the South Shetland Islands recounted that '[t]he smaller [elephant seals] were killed with lances, but [...] "brown cows" [that is to say, the adult individuals] could not easily be killed that way and were accordingly shot through the head' (Ames 1830: 143). Capt. William Robinson on Heard Island recalls that

> A cow elephant, when it just comes ashore, will yield nearly, or quite, two barrels. The March bulls that haul, as their name implies, early in March will often make five barrels of oil. They are not so easily handled as the cows, and are mostly shot in the roof of the mouth, while a very short lance is used for the cows.
>
> *(Nash 2009: 82)*

The use of lances and firearms was not the problem it would be if used on fur seals because it was the blubber of the elephant seal rather than the skin that was of value. The lance is variously reported as being 6–15 feet long (3.6–4.5 m) in length, with an iron head 24–39 inches (61–99 cm) in length (Delano 1817: 288; Cumpston 1973: 59; Busch 1980a: 122–123; Richards and Knight 1996: 21). A lance 4 m long with a rolled iron tip 16.5 cm long, riveted onto the shaft, was found outside a sealers' cave site on Rugged Island in the South Shetlands in 2005 (Pearson and Stehberg 2006: 341) (Figure 4.7), and others were found in the 1950s excavations of Lima-Lima Cave and Cutler 1. Photographs of these indicate broad diamond-shaped heads, and what look like whaling lance shafts (Lewis-Smith and Simpson 1987).

Evidence of firearms has been found at a number of sites (Figure 4.8). In the 1990s, the Argentine project recovered a gunflint at Lima-Lima and firearm projectiles at Cerro Negro. These balls of 16 mm in diameter were probably

FIGURE 4.7 Elephant seal lance: (a) Archaeological artefact found outside the Rugged Island 1 cave site (the models are Michael Pearson and Rubén Stehberg) (Photo by Michael Pearson 2005); (b) Detail of the rolled iron lance head riveted into the shaft (Photo by Michael Pearson 2005).

shot from muskets. In the 2010s, lead balls were recovered at the site Punta X2, having a similar size to the ones found at Cerro Negro. Lead balls of 3.5–4 mm in diameter were also found at Sealer 4 (Nolasco 2018b), probably being shotgun pellets. Pearson and Stehberg (2006: 344) reported the finding of copper percussion caps at the Rugged Island 2 site (common from the late 1830s).

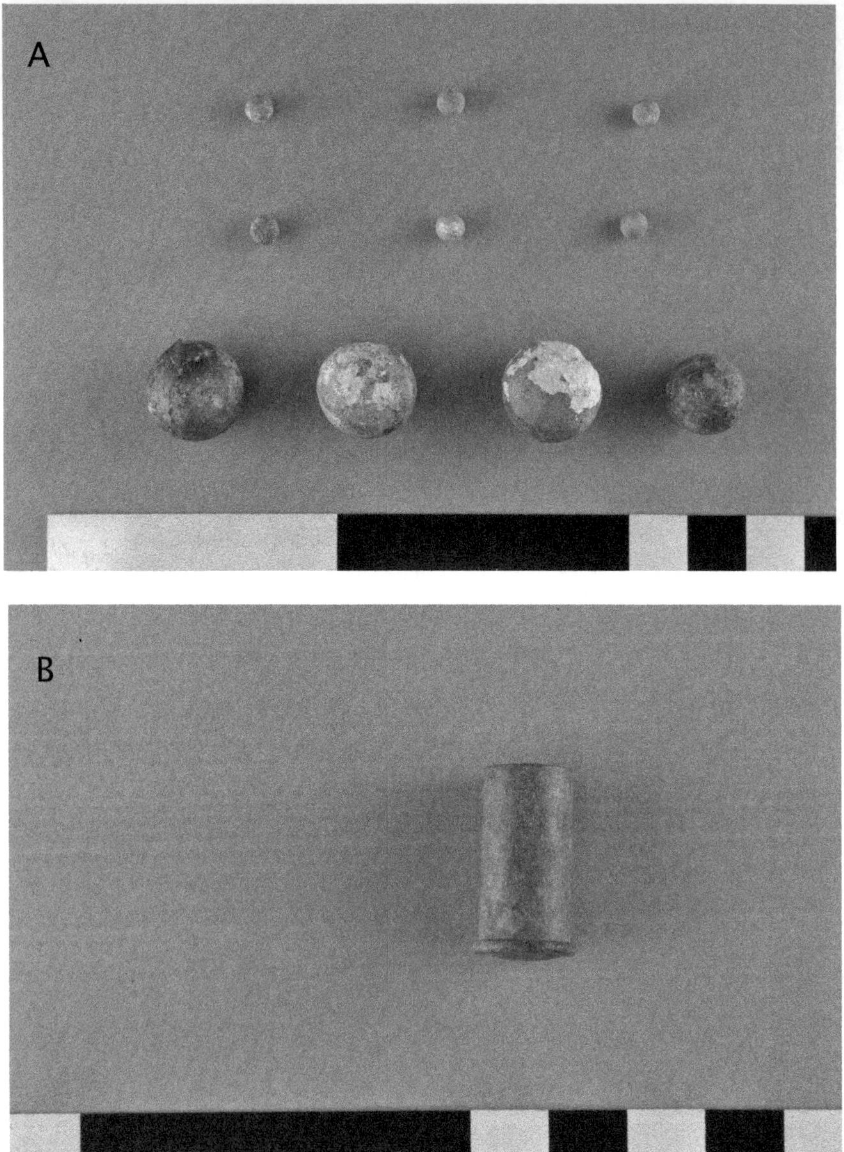

FIGURE 4.8 Ammunitions found at sealing sites: (a) Collection of lead ammunition of varying sizes from Punta X2 and Sealer 4 (Photo by LEACH 2021); (b) Copper percussion cap found at Sealer 4 (Photo by LEACH 2021).

The blubber was either tried out immediately in shore-based try-works, or sent out to a shallop in barrels or as floating strips, to be taken back to the ship or a central try-works ashore for trying-out. If the blubber could not be dealt with immediately, it could be buried or put into barrels to prevent the birds from eating it. Knives, including a curved flencing knife (one of which with a portion of its wooden handle was found archaeologically in the Byers Peninsula – Zarankin and Senatore 2007: fig. 155), were used for cutting blubber into large squares after the skin had been removed (Figure 4.9). This required turning over the huge carcass. The large pieces of blubber removed from the carcass were then sliced into strips about 1.5 foot long by 1 foot wide (45 × 30 cm), washed to remove any sand that would promote burning on the bottom of the try-pot, and these smaller portions cut part-way through with a two-handled blubber knife on a wooden bench or 'horse', then fed into the try-pot (Clarke 1850: 27; Clarke 1854; Phelps 1871; Richards and Knight 1996; Nash 2009).

Try-pots were large cast iron boilers, usually erected in pairs, with the rendered oil passing from one pot to the other by way of an attached lip or by using a copper dipper or ladle. Centralised try-works could also have a copper or iron tank located next to the try-pots into which oil was transferred for cooling (as

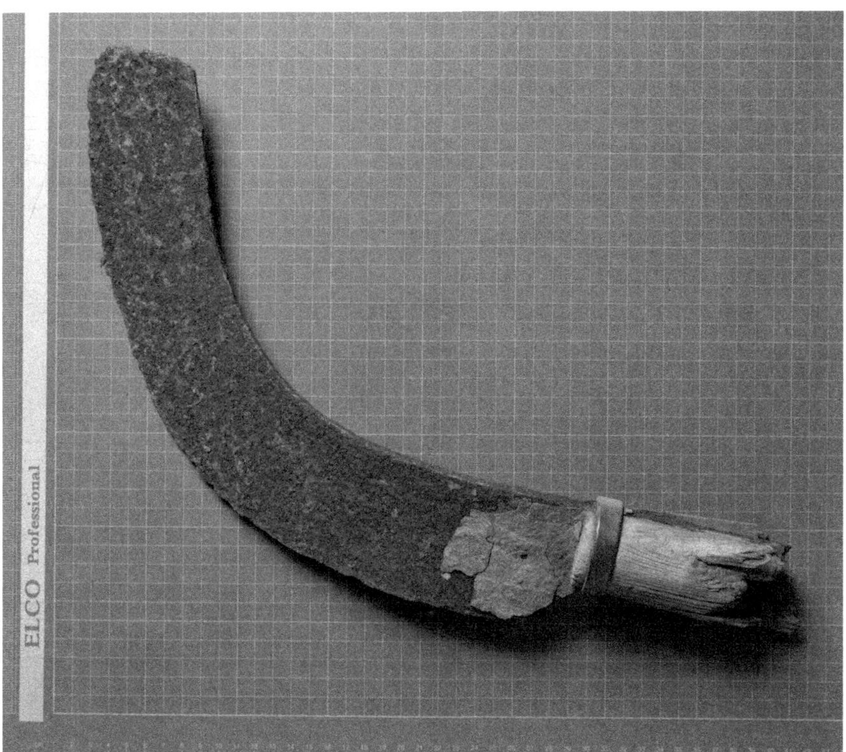

FIGURE 4.9 Flencing knife blade (Photo by Melisa A. Salerno 2021).

at Heard Island – see Chapter 2). Two examples of a blubber press, a wooden box with a screw-mounted plate at one end and a perforated plate at the other used to squeeze the remaining oil from fried blubber scrap, have been found at Heard Island and Kerguelen Island (the Heard Island example in the Tasmanian Museum and Art Gallery, Hobart, and the Kerguelen example in the collection of TAAF – Laetitia Therond pers. comm.). The scraps left after the oil was removed could be used as fuel to heat the try-pots. Once cooled, the oil was stored in casks. On some occasions, try-pots were placed on the islands and sealers had to find a suitable way for transporting the casks on board (for instance, resorting to a shallop – see above). A variety of tools associated with try-works have been found at sites, including copper perforated skimmers for removing scrap, forks and hooks for handling slippery blubber, flencing tools, grindstones for sharpening blades and barrel components (e.g., Lazer and McGowan 1990; McGowan 2000).

As mentioned in Chapter 3, several try-work sites have been located in the South Shetlands, at Cutler Beach on Livingston Island and at Flat Top Head on King George Island, consisting of rings of burnt ash and concreted material (Figure 4.10). Try-pots were expensive pieces of equipment, and were usually removed from land-based try-works when the season ended, leaving only stone heath bases or concreted oil to mark their former locations. However, a broken try-pot sits on the beach at Yankee Harbour, on Greenwich Island (Figure 4.11), with a try-work area nearby, and as described in Chapter 2, there are many try-pots remaining on sub-Antarctic islands.

FIGURE 4.10 Broken try-pot and try-works site in the distance, Yankee Harbour, Greenwich Island (Photo by Michael Pearson 2008).

FIGURE 4.11 Try-works site on Fildes Peninsula, King George Island. A try-pot and fireplace would have sat in each of the two stone rings (Photo by Michael Pearson 2007).

The remains of wooden casks have been found at different sealing sites. Lewis-Smith and Simpson (1987) dug up fragments of barrel staves and iron hoops, including examples near the more recently identified try-work site at Cutler Beach. Barrel heads and staves have been recovered at the Cerro Negro, Punta Elefante X1, Punta X1 and Punta X2 sites, and while these remains have not been thoroughly studied as yet, it is clear that they represented casks of different sizes and volumes (Figure 4.12). The presence of casks can also be interpreted from the archaeological finding of iron hoops. Wooden barrels and casks could have served different purposes. While barrels of sea water were used as ballast in large ships, the casks on smaller sealers are far more likely to have been filled with fresh water. Small casks were used to store perishable goods, such as biscuits, flour, spirits, dried vegetables and spare clothes, so the smaller casks found at sealer sites could have contained provisions, equipment or water.

Barrels for whale oil, and presumably for elephant seal oil as well, were mostly carried broken down into staves, head and hoops and built by a cooper either on board ship or ashore as needed. Fildes, for example, refers to the 'Cooper's grave' at Cora Cove on Desolation Island in 1820. References from the whaling industry point out that oil could have been stored in casks ranging in size from a few gallons to a barrel or a hogshead. While measurements varied, both within and between American and British units depending on the industry it was used in, a barrel in British usage might contain 120–160 litres (about 30–35 imperial gallons), while a hogshead held 240–300 litres (about 54–66 imperial gallons). Wooden barrels were sometimes used to store seal skins (Clark 1887: 238) and elephant seal blubber (Nash 2009: 82). Therefore, analysis of the residues found on archaeological casks still needs to be done to determine their possible content.

FIGURE 4.12 Barrel head from the Cerro Negro site (Photo by Melisa A. Salerno 2021).

Sealers took the skins of seals and the bubbler of elephant seals, leaving animal carcasses on the shores. Documentary records describe that sealers often found large amounts of bones on some beaches, bearing testimony to the activity of other sealers in previous seasons. Archaeologists have monitored the taphonomic agents acting upon the bodies of dead seals on the beaches of Byers Peninsula, finding out that after a period of approximately seven years the material traces of the animals cannot be longer detected as a result of the impacts of weathering, erosion, animal movement, etc. (Zarankin and Senatore 2007). This probably accounts for a general absence of the archaeological remains of the slaughters on the beaches. Archaeologists have nonetheless recovered animal bones inside sealers' shelters. Sebastián Muñoz (1997, 2000) conducted zooarchaeological studies on the remains that the Argentine project recovered in the 1990s at several sites on Byers Peninsula (including those collected during excavations at Lima-Lima, and surface collections at Cutler 1, Punta Vietor 1, Punta Vietor 2 and Sealer 3). He observed that the bones recovered inside the shelters were mainly the result of anthropic accumulation and were in large part from pinnipeds. Moreover, he argued that some of the marks identified on the remains of local fauna could have been associated with the processing of seal skins (for more references on the work of Muñoz, see below the section on eating practices).

Isabel Cartajena and Rafael Labarca (2007) analysed bone assemblages from sealers' shelters that were excavated by Stehberg (2003) at Yámana Beach, on Cape Shirreff. The researchers argued that these remains mainly resulted from a processing mode that involved a systematic, repetitive and highly efficient way of exploitation, oriented towards the optimisation of time, energy and materials. They claimed that this pattern could be observed in the recurrence of the location and disposition of cutting marks mainly associated with skinning and to a lesser extent with the disarticulation of the animal and meat removal. In line with ethnographic and archaeological information, skinning marks apparently concentrated on the distal section of the front and rear extremities of seal remains (i.e. the cutting around the flippers). Furthermore, the researchers established that cutting marks were probably produced by sharp metal tools consistent with knives.

The question of whether the archaeological sites were associated with the exclusive exploitation of fur seals or elephant seals; or if they could have been oriented towards a mixed exploitation, is difficult to answer in many cases. In a preliminary work, Senatore (2018) has suggested that the presence of objects with specific uses (like wooden pegs), the remains of animal material and processing marks on bones, and the specific use of some structures (including storage rooms and large hearths) could provide clues on the subject. Considering this, the researcher has suggested that in the case of 15 different sites she had the opportunity to analyse (mainly through published information), she could tentatively distinguish nine locations connected with the provision of seal skins and six with the preparation of oil. Moreover, she argued that three of the sites could have been oriented towards a mixed production. Further research on this interesting line of inquiry would need a more intensive analysis of existing evidence, and further targeted excavations to focus on the specific question. While elephant sealing may not have been the sole or part objective at a particular site, the evidence at many sites shows that elephant seals were killed for blubber for domestic fuel and leather for footwear. This complicates the analysis of the archaeological remains in trying to determine the relative concentration of fur or elephant sealing at a particular site, by bringing elephant seal evidence into sites that may have been otherwise totally focused on fur seal exploitation.

Considering the dialogue between archaeological and documentary evidence, and discussions raised by contemporary social theory, some researchers have started to analyse and interpret the relationship between sealers and the animals they exploited. Historical sources point out that sealers made detailed observations of animal behaviour in order to understand when was the best time to catch the largest possible number of animals. They generally took notice of, and shared information on which month of the year animals arrived at specific beaches, in which order and for what purpose different groups of individuals landed (for instance, females with pups, adult males, young males, for weening, mating, moulting, etc.). While gathering this information, indiscriminate opportunistic killing was still the norm, skewing the demographics of colonies and hastening local extinction. Sealers also attended to possible changes in animal behaviour, observing that seals could develop specific tactics to avoid or escape from the presence of humans as time went by. This could have had a clear impact on hunting and colony survival.

Ames (1830: 143) pointed out that when he first visited the South Shetland Islands: 'The seals [...] were so tame, that when hauling our boats on shore, it was frequently necessary to kick them out of the way'. Some years later, Weddell (1827: 140) recalled the same idea, stressing that 'when these Shetland seals were first visited, they had no apprehension of danger, from meeting men; in fact, they would lie still while their neighbours were killed, and skinned'. Notwithstanding, Ames (1830: 148) argued that at the end of his visit to the archipelago animals had become scarce and elusive. In a similar vein, Weddell (1827: 140) insisted that as the seals recognised the presence of humans as dangerous, they started choosing rocky locations as resting places, from which they could jump into water. Several decades later, Clark (1887: 429) understood that the scarcity and the new behaviours of the seals made the hunters change their killing techniques from the use of wooden clubs at a short distance, to the use of firearms at a long distance. In the case of elephant seals, Weddell (1827: 136) simply noted that 'when lying on the shore, and threatened with death, will often make no effort to escape into the water, but lie still...'.

Referencing phenomenology and the paradigm of embodiment (sensu Csordas 1990; Merleau-Ponty 1993), one of the authors has pointed out that in order to become a sealer, the men had to acquire certain degree of familiarity with killing and processing techniques (Salerno 2011). This implied learning how to use one's own body to deal with the materiality of specific tools, approach the animals and synchronise one's movements with the rest of the gang, among others. Here, it is worth providing some examples. Experience gave sealers the opportunity to understand how hard and on which part of its body an animal should be hit to be knocked down. Furthermore, it was only through repetition that a man eventually acquired the skill necessary to skin a significant number of animals per day. According to Scammon (Marine Mammalia 153, quoted in Clark 1887: 429):

> Some sealers became great experts in skinning the animals, and the number of skins one would take off in the course of an hour would be a decidedly fishy story to tell. However, to flay fifty seals in a day would be regarded as good work.

All of this required the sealer to transform the materiality of clubs and knives into some sort of extension of his own body. As previously noted, knockdowns were collective performances. Therefore, with a view to increase effectiveness, each man had to learn which was his role in the action, what to do and exactly when, how to anticipate the movement of others and the seals, understanding how fast and from which position it was wise to approach animals in order to avoid being knocked by them and to prevent them from fleeting.

Salerno has mentioned that hunting and processing techniques could have had a sensory-affective impact on sealers too; especially in the case of the newly exploited rookies. Whaling narratives describe the first impression caused by the bloodshed, the permanent contact with greasy surfaces, and the smoke stemming from try-works. These initial experiences could have created long-lasting memories, although repetition could have naturalised similar circumstances. Ames (1830) shared some reflections on his initial experience of sealing on the South Shetland Islands. On the one hand, he recalled that when he caught his first seal:

> I had knocked him down and was in the act of skinning him, when being only stunned and recovered by the pain of the operation, he uttered a piercing shriek, so exactly like a woman or child, that I leaped full two feet into the air, an exploit that I never performed before and probably never shall again.
>
> *(Ames 1830: 133)*

On the other hand, he commented that

> I saw one beach where upwards of three hundred had been killed, almost every one of which had a young one, as black as coal, and looking at a little distance, like black water spaniels. These poor little wretches were standing whimpering and whining each one by the mangled carcass of its dead mother, a piteous spectacle.
>
> *(Ames 1830: 144)*

From the standpoint of symmetrical archaeology (sensu González-Ruibal 2007; Witmore 2007), Raquel Nolasco (2018a, 2018b) has studied some aspects of sealing on the South Shetland Islands. Overall, her work has considered that studies on the exploitation of sea mammals have tended to be asymmetrical and anthropocentric, assuming linear and one-way relationships where only humans are understood as agents (while other participant entities are considered to be passive). Resorting to a broadened concept of agency (sensu Latour 1994, 2012; Dürbeck et al. 2015), Nolasco claims that the study of sealing exploitation needs to address the multiple flow of relationships among human beings, the animals people interacted with and the objects they used in hunting and processing. Therefore, she argues that approaching animals as agents implies understanding them as subjects conscious of the environment and capable of responding in different ways to the actions in which they participate. Meanwhile, objects give opportunities and impose constraints to human–prey interaction, providing clues on past actions. Considering archaeological and documentary evidence, Nolasco (2018a, 2018b) has explored how sealers, animals and tools shaped each other through their mutual interactions, involving gestures and more or less lasting contacts that could have changed their own materiality.

4.2 Subsistence and leisure time practices

The provision of food, clothing, equipment and shelter essential to the subsistence of sealers ashore consisted of a mixture of materials brought with them from far distant ports in their ships and those obtained locally on the beaches. The sealers' capacity to survive in the extreme isolation and harsh conditions relied largely on their ability to modify traditional activities and practices, or invent new ones that made up for the scarcity of imported goods and the necessity to live off the land.

The work of sealing and of ensuring their survival took up the majority of the time of those ashore. However, opportunities for leisure time did exist: such as when seals were not present at the end of a season, or had been thoroughly harvested, when the weather was too bad to work, and during the necessary downtime at the end of a working day. Leisure time involved both communal and individual activities that both reinforced group unity and allowed for quiet personal time. The archaeological evidence helps us identify some of the leisure practices.

4.2.1 Clothing

The clothing of the sealers in the South Shetland Islands was predominantly that of the merchant seaman of the time. We will look at this first, and then consider those aspects of clothing specific to Antarctic sealing. Once the men were hired for a whaling or sealing voyage and the seaman's articles (a sort of contract that stipulated the length and destination of the voyage, as well as the share or lay that each one would receive) were signed at the port of departure, the men were given some money in advance by sealing companies to buy an outfit. Outfitting stores were sometimes owned by whaling and sealing companies, representing a profitable business that allowed money given in advance to rapidly find its way back to the owners of the vessels. Furthermore, whalers and sealers frequently complained about the poor quality of the articles sold at these stores and their inflated prices. The most frequent victims of outfitters were the rookies, the 'green hands' who did not know what kind of clothing was appropriate for the voyage or the final destination (Holmes 1861; Nordhoff 1856; Davis 1874).

The *Whalemen's Shipping List and Merchants' Transcript 1843–1914* (a specialised journal devoted to the operations of the American whaling and sealing fleet from the mid-nineteenth century onwards) provides insight into the existence of numerous stores which were devoted to filling seamen's needs in the United States (Figure 4.13). Some of the sale advertisements in this journal suggest that at least in the second half of the nineteenth century the supply of goods could have been relatively standardised, as most of the outfitting stores offered more or less the same types of clothes (Salerno 2011, 2015). There was a general tendency to sell ready-to-use garments, even though some stores also offered custom goods. Depending on the source, there were variations in price and quality.

A variety of documentary sources indicate that US sealers' outfits – for instance – included different styles of pants, frocks, and jackets made in dark woollen, canvas, duck and waterproof fabrics; drawers, stockings and shirts (either plain or striped and in different colours) manufactured in woollen, cotton and linen fibres; shoes, boots and pumps; and different types of hats and caps. A seaman's outfit also included sail and other types of needles for the repair of clothing. While a similar source of information for British 'slopsellers' has not yet been studied, it is likely that British seaman clothing did not differ substantially from the American version.

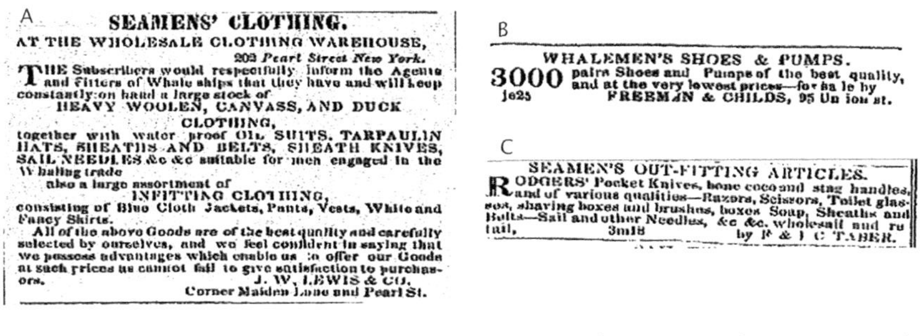

FIGURE 4.13 Advertisements for seamen's clothing outfitters: (a) References to woollen, canvas and duck clothing (Source: Whalemen's Shipping List and Merchants' Transcript – WSL 06/06/1846: 56); (b) Deals on shoes and pumps (Source: WSL 08/11/1847: 90); (c) References to personal objects other than clothes making up sailors' equipment (Source: WSL 05/07/1844: 32).

On board the vessels, hierarchical differences among the members of the crew are likely to have been marked and reinforced through dress, as the captains, the officers and the seamen had access to a different range and quality of products and probably wore them in different ways (Salerno 2011, 2015). The daily activities of a seaman working on a ship at sea produced many sources of wear and tear of clothing. The well-prepared ones may have had changes of clothing in their kit bags or sea chests – the ill-prepared and inexperienced may have very soon faced a clothing crisis. All seamen spent some of their leisure time washing and repairing clothing, skills that had to be learnt by each. It was said at the time that:

> Among whalemen, who perform all tailoring operations for themselves, necessity has brought in vogue a fashion called 'patch upon patch, and a patch over all'; and to such an extent does this prevail that it is said among sailors 'you may know a whaleman by his patched shirt'.
>
> *(Nordhoff 1856: 75)*

This expression could have been also true for sealers.

The possibilities of acquiring new clothes during a sealing voyage were limited. Some articles may have been bought or exchanged in intermediate ports or stopping points. Seamen's journals frequently indicate that on board ship the members of the crew had the additional opportunity to resort to the 'slop chest', a stock of clothes (and other personal requisites such as tobacco), which were bought by the ship-owners at the ports of departure to be subsequently sold to the men at sea (Olmsted 1841). As these products were sold at high prices that were later deducted from the lays, the seamen did the best that they could to avoid purchasing them. The American sealer William Phelps on Marion Island in 1819 recalled that aboard ship the slops store opened on Sunday mornings, while a healthy barter system operated among the men – one man hoarding his ship-biscuits and selling them in part for clothing. Phelps also refers to the sale of the clothing of seamen killed at sea (Phelps 1871: 36–38). Thomas Smith, leaving South Georgia in 1817 lamented that

> On leaving the island I suffered much with the severity of the cold, which affected my feet to a great degree. I had neither shoes not stockings, having previously worn them out, and it was impossible to obtain any from the slop-chest, as it was empty, or from any of the crew.
>
> *(Smith 1844: 137)*

When sealers were put ashore at sealing camps they do not appear to have taken their full kit bags or sea chests with them, due to the limited capacity of the boats landing the men and gear. With no possibility of replacement, the cloths of both officers and men would have suffered equally in the cold, wet, salt-laden and sandy, rocky conditions, dirtied and worn by the effort and bloody reality of sealing and living in cramped smoky shelters. All those sealing ashore likely found a need to mend their clothes far more often than when on board ship. Garments and shoes were discarded only when they could no longer be worn, and worn-out fabrics were recycled. It would not be strange for the men to have shared these fragments with their fellow sealers.

This make-do and the utilisation of local resources for clothing obviously increased with the time spent ashore, and in the case of some sealing gangs marooned on sub-Antarctic islands whole suits of clothes were made from local animal skins (see below) (Clarke 1850: 125–128; Phelps 1871: 92). This utilisation of local resources to augment the clothing supply is one aspect of sealing that sets it apart from the normal experience of merchant seamen. Clothing remains found at sealers' camps in the South Shetlands include textile garments and leather shoes. Textile and leather fragments represent one of the material remains more abundantly found at the sites. Documentary and archaeological evidence gives a rough list of some of the clothing used ashore: heavy hemp or linen-fibre, cotton and woollen cloth trousers; leather belts; flannel shirts; knitted socks, gloves and jumpers/sweaters; cloth or leather waistcoats; bone and wooden buttons; leather knee-boots ('sea-boots') and shoes, moccasins (made on shore); oiled canvas sea jackets and hats ('sou'westers').

Salerno (2006, 2011) has contributed a formal study of the textile pieces recovered by the Argentine project during the first excavations conducted at Lima-Lima, Playa Sur 1 and Cerro Negro (Figure 4.14). Her work has included the analysis of textile fibres, fabrics, stitches and possible garments. Conservation techniques have been undertaken to clean the textiles and gain visual access to their features (Salerno and Rodríguez 2007; Salerno et al. 2007). This confirmed the presence of both knitted and woven fabrics among the remains, with most of the knitted pieces thought to have been part of handcrafted woollen sweaters, gloves, etc. (probably produced at home or at sea rather than being mass-produced). Most of the woven fabrics found at sealing sites were fragmentary, but through detailed study, it was possible to determine that a large proportion of them were parts of external clothing, including jackets and pants made from woollen and linen fibres. Only one complete woven garment, a jacket, was found at Playa Sur 1. Cotton, jute or linen-fibre textiles used for seamen's garments did not apparently differ greatly from the fabrics used to produce sails. The presence of similar fabric in the sails used as roof coverings at some sealers' shelters seems to suggest the idea that sailcloth could have provided material for clothing patches and probably some hand-made articles. Woven fragments often show signs of repair such as stitches and patches made out of different fabrics, and the shape of some woven remains seems to indicate that they were cut from larger pieces.

Shoes proved to be particularly susceptible to wear and decay in the sealing environment. According to Ames (145–146), 'the very strongest of Lynn manufacture vanish[ed] from the feet like wet paper' – Lynn being a primary shoe manufacturing town in the United States. Salerno (2009) and Alkmim (2014) have studied the remains of leather shoes found at different sealing sites

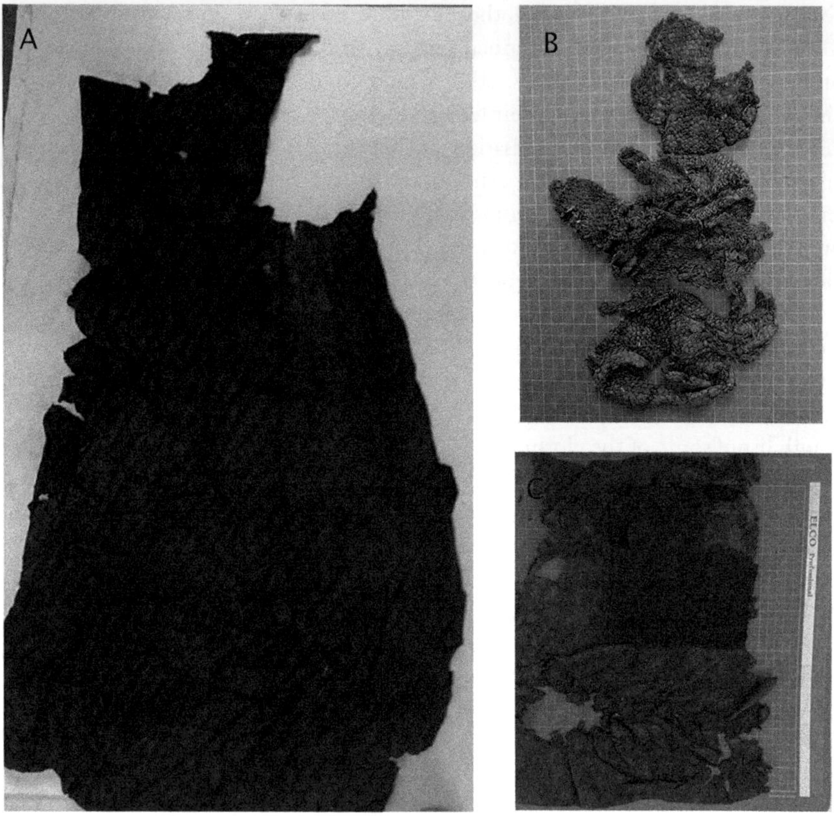

FIGURE 4.14 Archaeological clothing remains found on Byers Peninsula: (a) Front section of a jacket; (b) Woollen glove; c. Textile with patch made from another fabric (Photo by Melisa A. Salerno 2021).

(Figure 4.15). Small numbers of shoes have been found in several shelters, being well preserved due to the favourable conservation conditions for leather in the South Shetland Islands. However, as the threads that kept the pieces together suffered from decay, most shoe fragments were disarticulated. Whenever possible, the different components were reassembled to identify the style and the construction techniques of the articles, helped by the fact that the vast majority of the disarticulated components were found intact, with matching components being found separated in the archaeological deposit. The shoes represented standardised work shoe styles common in the period, most of them being low shoes with an upper section made up of two pieces (including the vamp and the quarters) similar to that of slippers and leather soles. Variations mainly responded to the differentiation in the sole axis and the shape of the tips but included some lace-up shoes and an isolated wooden sole. The technique used for attaching the upper and the lower section of the shoe was a hand-made welt. Both the inners and the outer sole presented a channel on the bottom side to hide the stitches, either to

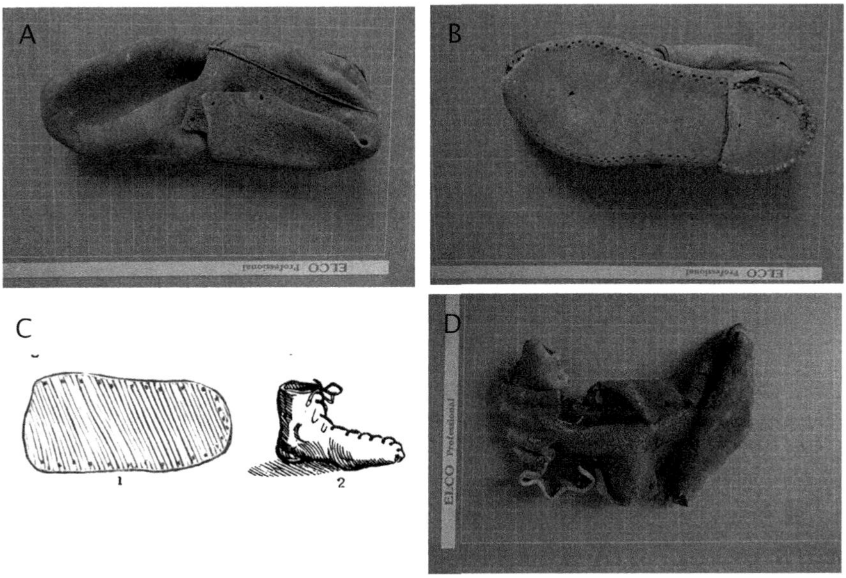

FIGURE 4.15 Sealers' shoes and moccasins: (a, b). Upper and sole of an archaeological shoe found at the Cerro Negro site (Photo by Melisa A. Salerno 2021); (c) Moccasin made from seal skin on Kerguelen in the 1820s (Source: Clarke 1850: 128); (d) Archaeological moccasin found at Punta Varadero (Photo by Melisa A. Salerno 2021).

protect the feet from the presence of the threads or to avoid the wear and tear of the stitches. The heels were fixed using wooden pegs.

The study identified signs of use, damage and repair. Almost all the shoes presented an intense wear of the channel that hid the stitches in the outer sole. Given this pattern, the men could have worn the shoes until the outer sole was practically detached. Some pieces showed severe signs of wear, holes on the soles and signs of repair on the heels (where iron nails were used to keep the different layers together). The presence of a significant number of leather off-cuts at sealers' sites probably reflects the attempts to reuse these shoes (Salerno 2009; Alkmim 2014).

When shoes could no longer be repaired, the next best thing was to make moccasins. These demanded little time or effort to make, as it was possible to

> put a foot on [a piece of leather] and cut around it, sew up the heel, and run a string round the toe, which draws it up, and tie it on the instep. By walking it becomes leathered and soft to the foot.
>
> *(Townsend in Clark 1887: 435)*

The collections studied by Salerno and Alkmin included several components used to make moccasins, manufactured in the camps. They consisted of pieces of leather that wrapped around the feet, having cuts on the edges so a lace could pull the leather together across the instep. While moccasins provided a poorer protection than shoes, they were easy and relatively quick to make, and they are found in a wide range of sealing shelter sites and in the literature for sealing in the sub-Antarctic islands. Moccasins made of 'cloth fabric' as well as ones made of skin, one example with an attached wooden sole, were reported from the 1959 collections made at Lima-Lima Cave and Cutler 1 site (Lewis-Smith and Simpson 1987). They are found in many other sites (as suggested in the studied collection); a whole example and many off-cuts, for example, being found in Rugged Island 1 site and at other sites on the Byers Peninsula (Stehberg 2003: 105; Pearson and Stehberg 2006: 341; Zarankin and Senatore 2007: 136–137).

The types of leather used to produce these moccasins have not yet been fully studied, but it is apparent from both the archaeological remains and the literature that both fur seal and elephant seal skins were used. Nathaniel Ames not only recounts that the skins of the old wigs (male fur seals) were not merchantable in Canton and 'we seldom killed them except to make moccasins of their hides' but also recounts taking enough skin from the back of what appears to have been an elephant seal (11 feet long) to make two pairs of moccasins (Ames 1830: 146, 148). Elephant seal skins had no market value. Charles Goodridge, speaking of his sealing and shipwreck in the Crozet Islands in 1820–21, writes that they used elephant seal skin for moccasins 'when we were out of the King Penguin's skins (see below), or wished to make a more durable article for wear; and these we used to sew together with thongs formed from their sinews' (Goodridge 1832: (1839 edition) 54).

William Phelps on Marion Island in 1818–20 denotes Sunday as the day they made moccasins:

> We found that common boots and shoes were entirely unfit for our shore-work, but that moccasins made of green hide of the seal of elephant, were just the thing. We cut a piece of fresh green skin oblong and oval, the size of the foot, with a suitable margin to punch holes in all around the edge; a string was rove into the holes, and the skin drawn up snugly over the foot, with the hair on the inside; with these we used dried grass instead of stockings, thus keeping our feet very comfortable. Whenever, in travelling, our feet became wet and cold, we had only to find a bunch of dried grass, and were all right again: or when a hole was worn in our moccasins, the nearest elephant furnished a pair, and it took but a few moments to transfer them from his neck to our feet.
>
> *(Phelps 1871: 65–66)*

Not yet found in the South Shetlands, moccasins made out of penguin skins are reported elsewhere. Capt Robinson recalled that on Heard Island:

> The men mostly wear shoes (or morgasons) made out of the beautiful skin of the king penguin. The skin, of course, is worn outside, the feathered

> side makes a warm, comfortable slipper. When any person requires new shoes he goes to the penguin rookery near by, and fits himself to what he requires.
>
> *(Nash 2009: 82)*

Charles Goodridge on the Crozets again writes that

> 'Of the skins of the King Penguins we used to form a sort of shoe, (called by South-seamen, Moccasins) by turning the feathers inward, and drawing them tight over our feet. They were very comfortable, but not very durable, for our excursions over the mountainous and rocky parts of the Island, would in a few hours wear them completely out.' Footnote: 'When the skins were very tender, half an hour's wear would suffice to bring them into holes, we used therefore generally to take two or three spare pairs with us on any long excursion'.
>
> *(Goodridge 1832; as reworded in the 4th edition 1841: 41)*

When sealers were shipwrecked and left for long periods before their rescue, the clothing issue became far more urgent (Figure 4.16). British sealer John Nunn commented that while his sealing gang remained on Desolation Island (Kerguelen) following the wrecking of their ship in 1825, their clothing wore out; in his own words, '[the garments] were getting so ragged and worn as scarcely to defend us from the inclemencies of the weather' (Clarke 1850: 128). In Nunn's case and that of William Phelps on Marion Island and Charles Goodridge on Iles Crozet, seal skins were used for the manufacture of clothing. In addition to the normal processes of treating skins as outlined above, in order to make the skins pliable and soft enough to wear, they rubbed them down with stones and sand. The sewing of such heavy material had to be invented on the spot: elephant seal sinew was used as thread, iron nails used as awls and albatross bones as needles. Buttons were carved from whalebone (Goodridge 1832: 39; Clarke 1850: 125–128; Phelps 1871: 92).

Salerno (2006) has discussed the role that clothing could have played in the construction of social identities among Antarctic sealers. Within this interpretative framework, she has discussed whether sealers' identities could have responded to the acceptance, reinterpretation or rejection of some of the principles of capitalism such as consumerism and social segmentation. Considering clothing choices, it is pointed out that while sealers could have access to some garments produced by the capitalist system, the complete acceptance of its fundamentals did not depend on the occasional acquisition of some of its goods. Moreover, sealers' commitment to capitalist principles could have been restricted. As an example, it is worth remembering that although ordinary sealers and officers may have purchased different articles, the shared experience of living and working on the South Shetlands would have left all participants dirty, oily and in distressed clothing, helping blur some social differences in appearance. In view

FIGURE 4.16 Sealers marooned on Kerguelen Island with clothing made of seal skins. Other equipment includes 'sou'wester' hats, elephant seal lance, seal clubs, knife case at hip and rifle (Source: Clarke 1850: 119).

of these and other ideas, Salerno has understood that sealers could have defined their own identities in a different way to those proposed by the dominant discourses of capitalism – clothing representing an active element for an alternative negotiation of how sealers perceived themselves and were perceived by others.

Salerno (2011) has also focused on the experience of dress among sealers with an aim to explore personhood among them. Personhood, the status of being a 'person', is tied to the relationships that each one establishes with society. Anthropological research has shown that the cultural understanding of the nature and boundaries of a person may vary on cultural grounds. Therefore, while some groups equate the notion of a person with individuals, some others recognise the existence of more relational entities. As the material dimension of personhood presupposes a body, and bodies usually present dressed during social interaction, then the dressed-body as a whole presents as a relevant avenue for analysis. Considering the subjective experience of one's own dressed body, and the inter-subjective experience of the dressed body of the others, Salerno has discussed the production and reproduction of an extended person among sealers, defying the conception of individualism (often assumed as the only form of being in capitalism). Therefore, her work refers to the experience of an 'esprit de corps' that embraced the men and even the vessel.

4.2.2 Eating

In the ports of departure, wholesale stores provided vessels with the provisions and equipment necessary for the journeys. As in the case of clothing, sale advertisements in the *Whalemen's Shipping List and Merchants' Transcript 1843–1914* describe the food supplied to sealing and whaling companies at least since the second half of the nineteenth century in the United States. The analysis of these and other documentary sources suggest the existence of some degree of standardisation in the type of food products that were taken on board. Sealing companies frequently bought large quantities of industrially produced articles (including salt meat, sea biscuits, etc.), as they represented ready-to-eat meals which could be preserved for a long time. Fresh food (such as apples, lemons, strawberries, plums and cheese) and different ingredients (flour, sugar, rice, salt, vinegar, beans, etc.) were also bought in American ports (Cruz 2014). Products of different qualities and prices would have been chosen by different companies or with an aim to attend the diet of different ranks on the vessels.

At sea differences in the ways that meals were prepared, served and consumed reflected the hierarchical differences aboard the ship. The sealing captain decided which barrels of provisions should be broken out at different times throughout the voyage. They also decided which was the serving size of the food that would be provided to the crew, although occasionally the crew could respectfully approach the captain asking for more (sometimes successfully). On board the vessels, the preparation of the meals was the responsibility of the cook, working at the galley where he had a stove and a cabinet for supplies, while the rest of the men were supposed to attend to other duties. The cook may have prepared specific dishes for the captain, the officers and the men (including different types, qualities and amounts of products) (Salerno et al. 2019). In some cases sealing vessels carried a steward to act as a servant to the officers (e.g., Townsend in Clark 1887: 435), and this person may have been responsible for selecting and preparing food for the captain and officers. Meals were served at specific times, helping to organise the daily routine at sea. Each rank ate at different places. While the captain and the officers ate at the cabin, the rest of the men ate at the forecastle (Nordhoff 1856).

First-hand accounts list common rations for seamen at different periods. William Phelps, as a 15-year-old and always hungry cabin boy on a sealer in 1817, recalls how bread (biscuits) was served out every Sunday, six pounds per man, while beef, pork and other perishables were served out each day, as per a bill of fare posted up by the captain. In the six-man mess (there being three messes on the ship), the meat portions were served out 'blind', but it was rigged so the green hands got the poorest parts, and Phelps felt that they were generally not allowed sufficient food (Phelps 1871: 26–27). They also had flour duff (a stiff flour pudding, perhaps with dried fruit, boiled in a cloth bag or steamed) two days a week, and rum could be exchanged for an extra serve. As they approached the Malvinas/Falkland Islands, the ration had been cut by a pound of bread with the substitution of a potato. The men approached the captain and got an extra half a potato each. Some men exchanged their daily gill (118 ml) of rum for two biscuits. Phelps ate 'slush', the left-over grease and scrapings from the cook's cooking cauldron on bread to make up for meat and once stole the captain's breakfast pudding, being beaten for it after confessing (Phelps 1871: 34, 39, 40–41). Other reports indicate ships that were better supplied: Cyrus Clarke (1854: 23) recalls lying-to off Livingston Island in a storm in 1853:

> The weather is severe, but we have a stove in the forecastle and plenty of coal and clothing, and food – thanks to Messrs. Smiths and Tom Brown, of Sag Harbor – and only on deck two hours at a time. This is one remove towards civilization on the part of our Captain.

The surgeon on an American sealer in 1851–53, Nathaniel Taylor, en route to Iles Kerguelen, records that ships were usually well supplied with 'codfish, beans, Indian meal, coffee, tea', as well as salt meat, though his ship had run perilously short of supplies due the miscalculation of the master of the expected duration of the voyage (Taylor 1929: 120). Sealing vessels also resorted to other locations to get food. For instance, the Azores were frequently visited to buy fresh fruit and vegetables (which were considered essential for preventing scurvy). Those who visited the coasts of Patagonia and Tierra del Fuego had a chance to hunt, gather and exchange some products with indigenous groups. In this way, the ordinary seamen ended up participating in the system of their own food supply (Salerno et al. 2018).

Historical documents suggest that the gangs often took to the sealing grounds a limited amount of supplies from the vessels, due to the small carrying capacity of the boats. It is clear that those spending more than a few days ashore were expected to take advantage of the food resources they could find on the beaches. In the case of the South Shetland Islands, Ames (1830: 132) pointed out that

> When a boat was sent away she was supplied with two barrels of bread, two or three pieces of clear pork, cut half round the hog, a frying pan, sheet iron kettle, a keg of molasses, another of 'Boston particular' [alcohol], 'lacking six days of being a week old' and a quantity of Boston coffee, videlicet rye.

A similar range of supplies is referred to at other hunting grounds. For instance, in the case of the Malvinas/Falkland Islands, Townsend (1797, in Clark 1887: 434) claims

> A sealing crew want a good stock of bread, molasses, and peas for coffee, and they can get along with little beef and pork, but to be out of bread or molasses for sweetening their coffee is very uncomfortable. They get very much attached to what they call slops, which is tea and coffee, in this cold uncheerful country.

Notwithstanding, historical sources point out that the men had to resort to local food resources to survive. Nathaniel Ames (1830: 159–160) recalled his trips in sealing boats in the South Shetland Islands in 1820:

> As for provisions, our bill of fare consisted of the liver, heart and flippers or fins of the seal, which parboiled and fried were excellent, 'elephants' tongues, far superior to neats' [cow], and several different species of birds. Eggs we had in plenty in the beginning of the season, so that we lived at least comfortably, and were always blessed with huge appetites.

Sealers traditionally ate local resources not only on the South Shetlands but also at other hunting grounds on the Southern Oceans. References to the use of specific parts of seals and elephant seals as food (including the heart, tongue, kidneys, flippers etc.), as well as birds and their eggs are often repeated. In the sub-Antarctic islands, a local cabbage-like vegetable was eaten, called Kerguelen cabbage (*Pringlea antiscorbutica*) or Macquarie Island cabbage (*Stilbocarpa polaris*) depending on the island. It was a bit unpleasant to eat by itself but was a powerful anti-scorbutic and was often used in soups or stews (e.g., Nash 2009: 87).

When unforeseen long stays occurred, food supplies could be a critical issue in survival, as William Phelps, left with a sealing party on Marion Island 1818–20 illustrates:

> We were on the island over two years; one year without bread or any other ship-stores, wholly without vegetables, and in fact, with nothing to eat but meat, eggs, and fish, with good cold water to drink … yet during the whole time I do not remember a single case of a man being laid by for one day with sickness….
>
> *(Phelps 1871: 91)*

John Nunn on Kerguelen illustrates the combination of local foods:

> We also removed the nose and flippers (feet) [of elephant seals]: of these we prepared a very palatable and excellent kind of soup, which we named 'flipper soup'. In the preparation of this we used first to scold the parts well, that we might clean them and remove the hair; after which we introduced them into fresh water and boiled them for some hours, occasionally adding slices of the leaves of the large cabbage-like plant which various parts of this island afford. By the addition of a little salt to our taste, we obtained not only a very nourishing kind of food, but truly a very palatable and agreeable dish, which we relished much.
>
> *(Clarke 1850: 130)*

Flipper soup was also prepared in the South Shetland Islands, described by Robert Fildes as 'not unlike mock turtle which it was humerously stiled (sic)' (Fildes 1821a: 64–65). The archaeological presence of numerous seal toe nails in sealing shelters in the South Shetlands may indicate the cooking of 'flipper soup', or their frying as described by Ames. In some cases, the use of seals by themselves as an anti-scorbutic was recommended for medical reasons:

> I will advise you, as a very necessary thing, to see that your men, at least those who are landed, eat plenty of seal meat, for it is necessary for their health. There is no danger of any man getting the scurvy if he will use seal meat freely, and as soon as they have used it a few days they will use no other meat, for it is first-class, and the pups that are three or four months old are as good as any pig.
>
> *(Athern in Clark 1887: 433)*

Nathaniel Ames seems to have had different taste sensitivities, as he described seal pup meat as: '…exceedingly delicate and well flavoured, much resembling lamb' (Ames 1830: 144). Robert Fildes agreed: 'The young pups make a good sea pye, not tasting the least of Blubber, the Sailors dignified their meat by the name of Pompey in lamb'. Fildes went on describing local recipes:

> [Sea Elephant's] tongues are excellent when corn^{d} and dryd. Their hearts when stuff^{d} and roasted are not inferior to a Bullocks: ... [Penguin's] eggs are very good particularly when fry^{d} or in pancakes ... The Penguin's and livers are pretty good eating and make a tolerable mess when there is nothing better to be had. ... Nellies [Giant Petrels] young used to supply us with many a good meal, being capital eating when curry^{d}.
>
> *(Fildes 1821a: 64–65)*

According to historical records, sealers prepared their food resorting to the few implements that they brought from the vessels, including a frying pan and kettle (Ames 1830: 159). Nunn (Clarke 1850: 109, 123–124) on Kerguelen Island recalled that: 'We always travelled, each with our three-pint hook-pot, called by the sailors a "no-favour," a tinder-box and matches: ... we also had a tin plate and a cup to drink out of ...'. Nunn mentions a particular way in which he and the members of his gang prepared their food, a procedure that might have been also followed in some other hunting grounds and contexts (Figure 4.17). They dug a hole into the earth in basin form, then rubbed a piece of rope yarn with bubbler, lighted it and placed it on the bottom of the hole. The burning rope yarn was subsequently covered with an iron hoop, which was bent in a zigzag manner. A piece of blubber was immediately placed on the iron hoop, the fire beneath melting the bubbler, causing it to drop and take fire. The frying pan or kettle was placed on three bolts previously driven into the ground forming a tripod.

The archaeological record in the South Shetlands confirms and expands on this approach to cooking (Figure 4.18): flattened hoop iron in fireplaces suggests their use as fire bars to support either blubber fuel or cooking vessels; iron trivets

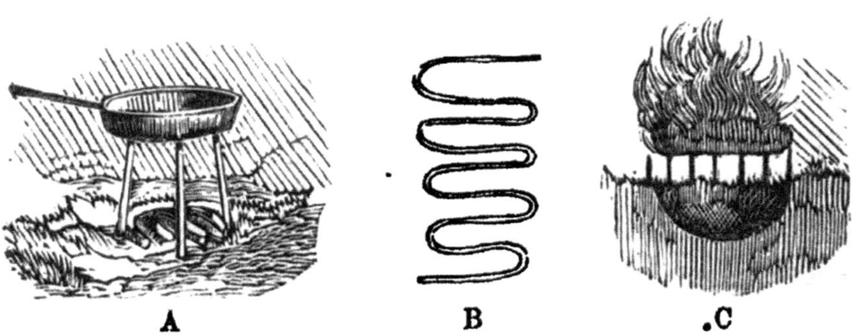

FIGURE 4.17 Cooking apparatus, as used on Kerguelen Island in the 1820s: (a) Frying pan on top of tripod of three bots; (b) Zig-zag of the hoop iron; (c) Blubber burning on the hoop iron (Source: Clarke 1850: 124).

were used to support cooking items above the fire in place of Nunn's iron spike tripod. The use of blubber as a fuel is well attested in the archaeological record, with heavy concretions of burnt blubber, often with charred elephant seal or whale bones in it which soaked up the oil and burnt well. Geochemical analysis has shown the presence of blubber residues spread throughout fireplaces in shelter sites (Villagran et al. 2013: 12).

At a mid- to late-nineteenth-century site on Rugged Island, a whole cast iron stove was excavated (Figure 4.19), and similar cast iron stoves and fragments have been found on Heard and Macquarie Islands (Burton and Williams 1985; Townrow 1989: 18; Pearson and Stehberg 2006). The remains of some other artefacts shed light on the restricted diversity of cooking and eating utensils at sealing sites: fragments of a broken cooking pot, one with three short legs, have been found (Figure 4.20); a kettle spout indicating a kettle to boil water and a variety of ceramics including transfer-printed cups (Stehberg 2003: 78; Pearson and Stehberg 2006; Zarankin and Senatore 2007: 133–143) (Figure 4.21). Zarankin and Senatore (1999, 2005, 2007) have suggested that the scarcity or absence of tableware remains at most sealing sites could reflect that they were not always taken ashore. Therefore, sealers could have simply eaten from pots and drunk from bottles, reinforcing the principles of sharing among the members of the gangs.

FIGURE 4.18 Fireplace at Punta Elefante 2 site, showing an iron cooking frame to right, hoop iron upper left and dense ash of burnt blubber and seal and whale bones. Whale bone upper right was a roof beam (Photo by Michael Pearson 2014).

FIGURE 4.19 Cast iron door of portable stove excavated at Rugged Island 2. The site dates from the mid- to late nineteenth century (Photo by Rubén Stehberg 2005).

FIGURE 4.20 Broken cooking pot from a sealers' shelter (Photo by Melisa A. Salerno 2011).

Muñoz (1997, 2000) undertook early research into some aspects of sealers' eating practices through the study of zooarchaeological remains recovered on Byers Peninsula. The researcher sought to understand the origin of bone accumulations at different sites identified by the Argentine project in the 1990s. The large quantities of food remains found in sealers' shelters, particularly around the hearths, were interpreted in the context where eating was part of a collective practice of gathering people of different ranks – there was neither room for, nor any evidence of social stratification and separation in living conditions within sealing gangs. As suggested above, the proportion of foreign and local resources consumed by a sealing gang varied depending on the supplies brought from the vessels and the length of the landings.

In general, the remains of foreign resources mainly consisted of beef and pork and to a lesser extent sheep and goat, as indicated by their bones (Muñoz 2000). Some of them presented butchering marks that suggested they were brought as standardised portions, probably preserved as salted meat. The remains of local resources included marine mammals and birds, and formed the majority of bone finds among the samples, especially represented by pinnipeds. In the case of seals, cutting marks and the presence of a disproportionate percentage of toenails suggest the consumption of the flippers eaten fried or in soups according to the journals, and an intensive use of the skulls, presumably to get access to the brain or tongue, noted in the literature as favoured parts. It was noted that the eating of viscera such as hearts, tongues, livers and brains, commonly attested to in the literature, would generally leave no archaeological footprint. The cutting marks

FIGURE 4.21 Artefacts excavated at Rugged Island 1 site. At the top is a British blue-and-white transfer cup in the 'Blind Man's Bluff in the Open' pattern of c. 1820 (Jon Prangnell pers. comm. 2007). Below are corks, a smoking pipe and the heel of a shoe (Photo by Rubén Stehberg 2005).

on most seal bones were interpreted as representing the skinning of the animals, rather than butchering specifically for eating, suggesting that parts of the seals killed and skinned on the beaches were carried back to the shelters for cooking. In the case of birds, cutting marks indicate the consumption of the animals' breasts and legs. Considering that choices were limited, it is probable that all the men would have had access to the same meals.

Taking into account that seals, elephant seals and penguins had been observed sitting on, laying near or inside the shelters, the question of whether bones were natural deposits or human food remains within sites remained to be answered. In his study, Muñoz determined that animal remains found inside sealers' shelters were the result of anthropic accumulation. Although there could have been overlaps between human and animal occupations, sealers' shelters presented a mixture of local and non-local animals, and remains demonstrating processing marks and a selective representation of anatomical parts (e.g. no fully articulated skeletons) that clearly indicated human agency.

María Jimena Cruz (2014, 2017) extended the research on sealers' eating practices. Previous work and new excavations conducted by the 'Landscapes in White' project allowed her to study larger bone collections from different sites on Byers Peninsula, such as Cerro Negro, Punta Varadero, Sealer 3, Pencas 3 and Punta Elefante 2. Determining animal species allowed distinguishing between local and non-local resources. In order to study taphonomic processes, Cruz attempted to identify body parts and how they might have been prepared, consumed and discarded. She argued that the proportion of local and non-local resources recovered at each particular site varied, depending – for instance – on the supplies brought from the vessels and the length of the landings. The foreign and local resources identified in the collections, as well as the cutting marks detected on the bones, echoed previous findings by Muñoz.

Venturing into the field of sensory archaeology, Cruz (2017) discussed the relationship between eating and memory, seeking to understand how sealers 'incorporated' a series of contexts with which they interacted. These contexts included the ports of departure and intermediate locations that were part of the sealing voyages, as well as the South Shetland hunting grounds. People's encounters with the world include sensory and emotional relationships where memories are produced and reproduced. This is how people create a sense of familiarity with specific places. As the relationship that people establish with the surrounding world brings into play their own body and the material dimension of things, then one possible way to approach the incorporation of a given context includes the study of eating.

Cruz argues that the incorporation of the South Shetland Islands could have depended on sealers' previous experiences in some other contexts. Food implied a specific way of relating to Antarctica, where the memories of what was eaten and how to do so mingled with new experiences. When the men visited a sealing location for some hours, the food was taken from the vessels. When they slept under up-turned boats, the first priority was killing an elephant seal for fuel, large blubber fires being kept going outside the boat camp (Pearson 2018a). In more lasting camps, the use of local resources gained importance and forced a greater participation by the men in their own survival and comfort. Eating food brought from the vessels could have made the men recall their lives in past scenarios, including their home towns or the ports of departure. Meanwhile, the incorporation of local resources and discovering ways to prepare them for eating might have introduced sealers to new flavours and experiences, helping them become acquainted with the Antarctic territory. While living effectively off the land did not make Antarctica 'home', it may have modified the first perception of it as unliveable barren wilderness.

Archaeologists have established comparisons between eating practices on board the vessels and on the South Shetlands grounds. While on the shipboard context provisioning, good or bad, was in the hands of the captain, and its preparation in the hands of the cook, the supply and the preparation of their own food ashore (labelled 'self-provisioning' by Maddison 2014: 112–117) could have been a more collective decision. These circumstances led to a closer relationship between sealers and food. Getting food from local resources required the participation of the gang in different actions, such as killing, chopping, and transporting meat. Furthermore, in the absence of a cook, all or at least some of the men engaged in the preparation of the meals. The manners of preparation were not necessarily standardised as on board the vessels. Eating on the islands was part of a collective practice, taking place in the undifferentiated space of the shelters (in particular, around the hearths where officers and the rest of the men gathered) (Cruz 2017). Considering that food choices were limited, everybody had equal access to the same meals.

Some people discovered freedom in the self-reliance ashore and the separation from the ship and its hierarchical controls. As an example of this, Nathaniel Ames wrote (1830: 159–60): 'In short we had become so attached to this roving, free and unconfined life, that a dozen or fourteen of us volunteered to stay all winter, a motion that was "over-ruled" by the captain'. On Más Afuera (Juan Fernández Islands), Ebenezer Townsend met a young English seaman working alone on the island. All he asked for was to sell some seal skins he collected for the re-filling of his small cask of rum:

> He left us, apparently as happy as a lord, saying he enjoyed himself better on the island than he ever did before in his life. He could sleep when he pleased, and work when he pleased; there was 'no starboard nor larboard watch,' and 'no one to quarrel with'.
>
> *(Townsend 1797 in Clark 1887: 464)*

4.2.3 Drinking

Drinking practices, in the intake of sustenance, can be defined as eating practices, although the act of drinking can also be associated with specific cultural practices and understandings beyond eating, as we will refer to later. This is the reason why we approach drinking in a separate section. Water was the most important commodity ashore and afloat. Without it, we die in a short period. Sealers undertaking boat trips and landing on shores normally carried a small amount of water in a small cask, to provide the necessary short-term drinking water for the crew. One of the critical characteristics determining the location of sealing shelters was the relatively close access to fresh water (Pearson and Stehberg 2011: 89). We have seen above that coffee and tea were staples listed as taken ashore by sealers, these being a necessary flavouring for what could otherwise have been stale or contaminated water, and fortuitously also requiring the boiling of the water.

The issuing of alcohol to sailors was a common practice in nineteenth-century naval and merchant ships. The daily 'grog', usually watered down rum, but able to be substituted for with other alcoholic beverages when available, was an expected part of the seaman's life and considered part of the 'payment' for the tough life and work (Maddison 2014: 137). Those who did not drink sold it on for other goods, such as additional food or clothing. Nunn records the use of alcohol as an additional reward for hard work launching and carrying stores to the shallops in Kerguelen Island:

> 'The captain promised us an extra allowance of grog when the shallops were safely launched and the business of the day was over'. After the day's work the men 'returned to the ship, and according to the captain's promise received their extra allowance of grog, with which they retired to their cabins and wiled away the evening in happiness and joviality, telling merry tales and drinking to their absent wives and sweethearts, a prosperous season in the whale and seal fishery, and a happy return to old England! The grog being finished and jokes exhausted, they turned into their hammocks and slept securely during the night'.
>
> *(Clarke 1850: 21–22)*

As mentioned above, Phelps not only recalls that even as a 15-year-old cabin boy he was issued with a gill (118 ml) of rum as grog a day, at noon with dinner, but also states that few of the crew, except for hardened 'old tars' drank it. When later shipwrecked, Phelps says that a demijohn of rum remained untouched for months until the cooper, the only drinker in the party, took it (Phelps 1871: 38, 70).

There is archaeological evidence to augment the written evidence of the use of alcohol in the Antarctic and other sealing islands. Ames (1830: 131) lists a keg of 'Boston particular', a raw alcohol, probably rum, being among the stores carried by a sealing boat heading off to the beaches in the South Shetlands. The remains of glass bottles and small casks have often been found inside sealers' shelters in the region. In the 1990s, Paula Moreno (2000) studied glass remains recovered by the Argentine project during the excavation of Lima-Lima and Playa Sur 1, and surface collections at Cutler 1, Punta Lair, Pencas 3, Punta Varadero, Punta Ocoa, Rocas Largas and Sealer 4. Thanks to reassemblage and the formal analysis of the artefacts, a minimal number of 20 bottles was represented. The bottoms were more numerous than other parts of the bottles, as they are heavier and more resistant to breakage.

The bottle types were dated between 1794 and 1820, consistent with the 1820s sealing period, and were English cylindrical black glass bottles (that were often used as alcoholic beverage containers). The bases of the bottles present both dome and conical shapes (with marks of sand pontil). The horizontal line in the union between the shoulder and the body confirms that the bottles were blown into deep moulds. These products were extremely popular at the time, as they were sold in every country having commercial relationships with the United Kingdom. Therefore, they cannot be used to determine the nationality of the people who occupied the sites. Some of the fragments showed incised and chipping marks that could have indicated the refilling of bottles and the use of broken pieces of glass as tools.

Fieldwork conducted by the 'Landscapes in White' project allowed Fernanda Soares, Luís da Rosa, Taísa Joia and Will Pena (2016) to approach glass fragments recovered at different sites on Livingston Island, including Sealer 4, Pencas 3, Punta Varadero, Punta X2, Punta Elefante 2, Punta Elefante X1, Rocas Largas and Punta Diablo. Among other things, the researchers reassembled the bottles, determined the minimal number of containers, made a formal description of the artefacts and conducted a distributional study of the remains at each site (Figure 4.22). Overall, they were able to reassemble 13 different bottles of black glass, of an English type and date range similar to those studied by Moreno (2000). All of them presented a tall cylindrical shape, except for one being a long-necked bottle. Cylindrical bottles were often used as beer or wine containers, while long-necked bottles were sometimes used for distilled beverages. A minimum number from one to four bottles was found at each site, and the remains were recovered at different areas of the camps – from sealers' shelters to small structures probably associated with work activities. Interestingly, none of the bottles were the square 'case' bottle form normally associated with gin, and often found in mainland excavations, suggesting gin may not have been a common alcoholic beverage favoured by sealers. Stoneware bottle fragments have also been found at South Shetlands sites, and these were also used, among other things, to contain beer and other beverages (see Stehberg 2003: 79).

Fernanda Soares and Gilberto Gardiman (2017) analysed the content of bottles as a clue to their use as alcohol containers, while bearing in mind that bottles could have been reused for other beverages taken ashore. The presence of granules of millet, barley and wheat and the marks of enzymatic action upon the granules suggested that the cylindrical bottles contained beer. The interpretation of the possible origin of the alcoholic beverages was not attempted, as no references of the particular range of grains and their blends used in American and British beers at the time was accessed by researchers. The long-necked bottle could have contained a distilled beverage such as rum. The findings made by Soares and Gardiman (2017) suggest that beer could have been more commonly consumed by sealers ashore than distilled beverages or wine. This could have

FIGURE 4.22 Reconstructed excavated bottles from sealers' shelters (Photo by LEACH 2021).

been associated with sealer's preferences or the types of alcohol provided by the owners of the vessels and the captains. It might also be that captains tried to limit the taking of spirits ashore, where its consumption could not be regulated as it was aboard ship. Beer, being much less alcoholic, might have been seen as a safer alternative. Rum is nonetheless known through the literature to have been transported ashore in small casks as well as bottles, but as yet no remains of small cask timbers have been identified in the sites.

These studies of alcohol consumption among sealers attempted to place them in the context of modernity, where conflicting trends influenced alcohol intake. Soares and Gardiman (2017) pointed out that at the end of the eighteenth century, the idea that an excessive alcohol consumption represented a vice or a disease became increasingly popular in the upper and growing middle classes. Within that framework, campaigns of moral reform were aimed at promoting docile and productive workers. Nevertheless, among working groups, alcoholic beverages were often thought as a way to step away from daily concerns and as a relevant dietary supplement. It was also a safer way to hydrate than provided by often dangerous or at least poor-tasting water supplies. In the nineteenth century, sealers participated in different scenarios where they could find more or less freedom for alcohol consumption, from seamen's taverns to life on board the vessels and life on the hunting grounds (Siqueira 2014).

Although alcohol rations were traditionally issued on board sealing vessels, the religiosity of some captains and their crew (particularly those from American evangelical communities) could have influenced tolerance to alcohol consumption. A number of sealer memoirs suggest that high levels of alcohol intake were limited to a few determined drinkers and that many others drank little or no alcohol (e.g., Phelps 1871: 39), although there was a market for it aboard ship, trading rum for food or clothing. It must also be remembered that these authors were writing for an audience that favoured abstention, so may have tailored their narratives with that in mind. Although the limited number of bottles found at sealers' sites could have responded to a bias in the discard versus the retention of bottles for reuse, or to a limited consumption of alcoholic beverages, the presence of the containers might also attest to the greater freedom of choice ashore.

4.2.4 Smoking

At the end of the eighteenth century and during the nineteenth century as a whole, pipe smoking of tobacco was a popular practice among all social classes, including sailors – and was equally opposed by others. Merchant vessels usually offered tobacco and pipes from the slop chest which were later deducted from sailors' pay at the end of the voyages. Historical records make reference to the practice of smoking at different sealing grounds. Moreover, they describe some of the innovative ways in which sealers managed to replace their pipes and tobacco when they lacked these products. Nunn (Clarke 1850: 131) pointed out that

> We had amongst our stores a small quantity of tobacco; but, having lost our pipes, we were at a loss to know what device to have recourse to for the purpose of forming a substitute. We were not long in removing the difficulty; for, by a little application of our inventive and mechanical powers, we constructed some of the canine teeth of an immature elephant seal, into the side of which we fixed the shaft of the wing-bone of an albatross!

Goodridge (1832: 53) and Phelps (1871: 69) also referred to the use of elephant seal teeth and bird bones to create pipes, and Goodridge added that 'Having no tobacco, [they] smoked the dried grass that grew on the island…'. The spread of these descriptions, on Kerguelen, the Crozets and the Prince Edward Islands, suggests that this resort to hand-made pipes was a common response in the 1820s.

The remains of smoking pipes have been found at different sites on the South Shetland Islands. Some 20 clay pipes have been excavated since the 1990s, and a metal pipe has also been found (Figure 4.23). Researchers have made a formal study of the pipes gathered by the 'Landscapes in White' project, considering

FIGURE 4.23 Collection of South Shetland smoking pipes (Photo by LEACH 2021).

their shapes, sizes, decorations and dates (Amorim 2019; Soares et al. 2019). Furthermore, they have applied 3D modelling techniques to the conservation of the artefacts (Brabo 2019). The collection included plain pipes (with no embossed motifs) and some others with standardised vegetal-floral patterns. Considering that clay pipes were fragile and they often had to be replaced, the presence of similar standardised patterns at the sealing camps could have been connected with the purchase of pipes at the vessels' slop chest.

Focusing on tobacco consumption, Fernanda Soares, Clarice Amorim and Lucas Pena (2019) have discussed the construction of gender identities among Antarctic sealers. Resorting to Butler's performativity (2000, 2003) and Connell's approach to hegemonic and subaltern masculinities (2005), the researchers have explored how smoking practices could have produced and reproduced certain ideals of masculinity, and how the materiality of the pipes could have played a specific role in that process. Overall, the researchers relate that in the nineteenth century, smoking was a predominantly masculine activity (although several women also engaged in the practice). Notwithstanding, during that time there were different prescriptions on how different sociocultural groups had to smoke. In the case of sailors, smoking could blur some social distinctions (including racial and ethnic differences) through the reiteration of a sense of belonging to a larger group: the community of seafarers (Pena 2013). This identity could have been constructed in contrast to other masculine identities not considered as tough as those of seamen (including the identities of the bourgeoisie, which were associated with a certain moderation in consumption, the aesthetic choice of specific pipes, etc.).

4.2.5 Some other leisure-time activities

Some other activities were carried out at sealers' shelters while the men were not hunting or processing seals or elephant seals (Hissa 2012). Historical sources mention that men of the fishery industries could have played several kinds of games of amusement both on board the vessels or on the hunting grounds. Although card playing could have been prohibited by some captains (as competition and gambling represented a potential source of tensions), the men could have find their own way to take cards and play games such as 'old sledge' and 'seven up' (Clark 1887: 438). Board games could have included back-gammon and some others (Goode 1887: 231). In the 1990s, the Argentine project recovered a hand-made board game at Playa Sur 1 (Figure 4.24). This was a piece of wood with a grid of lines carved on it, the boxes marked in alternate diagonal rows as in a chequer-board manner, with the same number of squares. Along with the wooden board were found round and squared game pieces, made with wood and leather. Some of the wooden pieces were marked with an X. Leather pieces from which the gaming pieces had been cut were also found.

Besides drinking, smoking and playing games, another form of entertainment among sealers included the creation of some simple works of scrimshaw (bearing in mind that more complex scrimshaw, if it existed, would have been taken home and not abandoned at camp). Scrimshaw describes the practice of carving of bone or ivory pieces with different motifs or into different objects. Whalemen

FIGURE 4.24 Wooden game board and leather pieces excavated at South Beach 1 site (Photo by Melisa A. Salerno 2019).

were known for the creation of elaborated designs on whales' teeth and for making objects out of whale bone and baleen. In the case of sealers, it is difficult to find documentary references on the subject. However, the presence of some cuts and perforations on the teeth of marine mammals found on the South Shetland Islands probably represents simple expressions of scrimshaw (Pearson and Stehberg 2006: 342–343; Pearson 2018c: 43–44) (Figure 4.25). The raw materials for

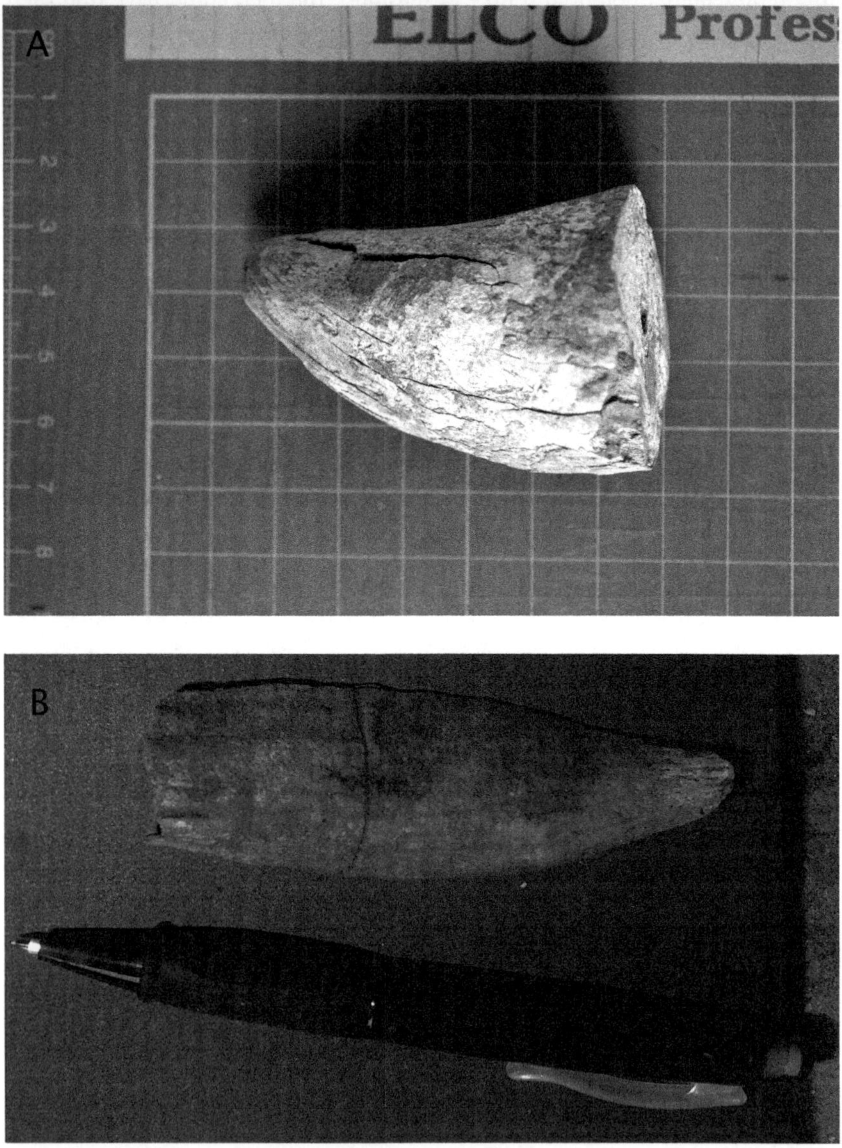

FIGURE 4.25 Cut and grooved elephant seal and sperm whale teeth from sealing sites: (a) (Photo by Melisa A. Salerno 2021); (b) (Photo by Michael Pearson 2005).

scrimshaw in the form of whale skeletons are commonly found on the beaches of the South Shetlands, and seal bones and teeth were obviously available. A more robust type of sculpture was found on an elephant sealing try-works site on Heard Island, in the form of a small basalt boulder on which a man's bearded face was carved in relief (Burton and Williams 1985).

Another pastime, walking and exploring glaciers, is not reflected in the archaeological record but was a common recreation when sealing season had finished or the local resource had been killed (Ames and Phelps quoted in Pearson 2018c:44).

4.3 Conclusion – everyday life scenarios

Based on the evidence discussed in the previous chapters, we can recreate to some extent the everyday life scenarios experienced by sealers. Historical documents have traditionally told us that sealing and whaling companies played a leading role in the early occupation and exploitation of Antarctica and the surrounding seas, and this was particularly true in the case of the South Shetland Islands. Documentary evidence gives us the summary outline of how sealing captains explored the islands and put sealing gangs ashore, with very few journals and memoirs by sealers themselves telling of their experiences. It is the archaeological evidence, however, that offers invaluable data expanding our knowledge of the sealers' daily life on this wild and previously uninhabited territory. It deals with the things the captains and owners were not that interested in, as the sealer's experiences were generally seen as just ephemeral detail about one of the inputs to the economic enterprise. Archaeological surveys and excavations have shown that a large number of sealing sites were established in the region and provide a particular insight into their occupation and organisation. Archaeology not only uncovers the details lost from the documentary record, but it allows us to go further, generating alternative lines of evidence on the diversity of the lives of these people involved, such as the form, variation in the size and spatial organisation of the campsites; the associated social organisation of the gangs; the particular uses of accommodation and work areas; the inadequacy of ship-based food supply and the need to rely on local food and fuel resources; and the material resources with which sealers had to survive Antarctic conditions.

Once on land, sealers created their own living spaces where most of their everyday activities took place. The characteristics of the shelters, such as the great variety in their design and relationship with work areas, suggest that decisions were made by a gang based on actual conditions, not on a pre-determined plan (Zarankin and Senatore 2007) – there was no prefabrication of kit huts as occurred in the later 'heroic era'. Given the small size of shelters and the lack of evidence of social stratification in their habitation, it is clear that the sealing gangs largely shared the hardships and comforts communally. Studies of the material culture from the sites reinforce the view that most activities conducted in the shelters probably involved social participation and a certain lack of hierarchies. Food remains showed that sealers shared mainly local resources, particularly seals, to augment inadequate food supplies landed with the gang: there is no evidence of segregation where some inhabitants lived on local food while others lived on food brought from the ship (Muñoz 2000, Cruz 2016, 2018). The absence of high-quality clothing and accessories among the generally low-quality textiles found in the shelters, and the substantial evidence for clothing repair and replacement with local resources (especially in the case of moccasins and shoes) suggests that the gangs shared a similar standard of clothing, as one might expect from a typical merchant marine cohort undertaking hard work (Salerno et al. 2007; Salerno 2009, 2011, 2015). The presence of games such as checkers, smoking pipes and basic scrimshaw work suggests a communal sharing of pastimes within or near the shelters (Stehberg 2003: 115; Pearson and Stehberg 2006: 342–343; Zarankin and Senatore 2007: 138–140).

The strategy for carrying out sealing varied greatly, as is documented in the historical literature. Some ships operated independently, dropping sealing gangs off at one or several locations, then sailed away to a safe anchorage until it returned to collect the gangs and their skins, blubber or oil. More often, it seems, mother ships moored in a safe harbour with a skeleton crew, and sent one or sometimes two small shallow-draught 'shallops' out to the sealing grounds, the shallops, in turn, delivering the gangs to the beaches in small whaleboats and returning at intervals to resupply the gangs and take off the skins, blubber or oil. A third strategy was to anchor the mother ship in a safe anchorage and send out gangs independently in whaleboats, the gangs returning when they had a load of seal product. This strategy seems to have been used in the very early period of sealing, when the location and scale of the resource were unknown, and again at the end of seasons and towards the end of the boom era when seals had become scarce and gangs needed to spend less time on beaches to collect them (Pearson 2018a, 2020). In all cases the number of men making up a gang seems to have been between 5 and 11: the first number being the crew of a single whaleboat, while the latter would have had to represent several landing to transport the crews (see Ames 1830; Burdick 1821; Davis 1822; Stackpole 1955 for examples). The number of men remaining on the mother ship was probably small in all strategies, as the ship was safely anchored and it was unproductive to have men idle. In some documented cases only one or two men were left aboard, resulting at least once in the loss of the ship in a gale (Fildes 1821a).

The time sealing gangs spent ashore on the beaches varied with the strategy adopted and with the number of seals available. Boat-based gangs might only spend a day or two on a beach, and live under their overturned boat at night; the boats carried too small a load to be able to supply a gang for a longer period or to carry back large amounts of sealing product. Nathaniel Ames records boat expeditions of up to 35 days resupplied and presumably relieved of their sealing products from time to time by the shallop (Ames 1830: 143, 152). In other strategies, gangs might spend weeks or even months in camp (as confirmed by the archaeological evidence). Those dropped directly from a ship appear to have been less frequently visited and resupplied than those serviced by shallops, to the point in some sub-Antarctic islands where the ship went whaling or on merchant voyages after dropping off a crew, and sometimes did not return for what stretched into years.

The nature of group dynamics and socialisation accompanying these sealing strategies is an interesting question. Aboard ship before deployment of the gangs and after they returned at the end of a season, life was strongly structured by discipline and social hierarchy – master, mates, petty officers, seamen. Different ranks served different functions and enjoyed different freedoms and comforts. However, as we have seen, the archaeology and some of the literature show that life in sealing camps appears to have been far less hierarchically structured and the sealers enjoyed greater freedoms, albeit within the overall imperative of gathering as much skins or oil as possible. The sealing gang was led by a mate or a petty officer and might have included a cooper if elephant seals were being harvested, but the rest of the crew were ordinary seamen/sealers. The group dynamic in camp was obviously different from that aboard ship, and the tenor of it must have depended on both the attitude of the mate or petty officer heading the gang, and the attitudes and behaviour of the gang itself. An example from the sub-Antarctic Marion Island is when the American sealer William Phelps was the youngest member of a gang left ashore sealing for a proposed nine months, which turned into two years. Quite strict discipline was maintained by the Executive Officer Enoch Burnham ('the Governor') to whom Phelps acted as 'Secretary': the two officers, Burnham and the Cooper, together with Phelps occupied one hut while the other four members of the boat's crew occupied a second hut. However, the crew gained significant freedom as the duration of the stay extended, working in small groups independently for much of the time (Phelps 1871: 54–80).

Where shallops were employed to supply the camps, there would have been relatively regular contact with the outside world (meaning the ship) and the chance existed to leave the beach if that was necessary. The example of the Davis and Burdick sealing gangs on the Byers Peninsula indicates the relatively frequent exchange of gang members in this instance. In other situations where the shallops visited less frequently, or where the mother ship was absent for a long time, the sealing gang would have felt entirely isolated, with the ever-present anxiety of the possible loss of the shallop or mother ship and their being marooned at the end of the earth with nobody knowing where they were (see Goodridge 1832: 19). The potential exists of finding archaeological evidence that indicates behaviour related to this anxiety of isolation, which has characteristics very similar to those experienced by shipwreck survivors, as discussed in Chapter 2.

As we have shown, sealing structures were primarily created with local resources. Other than a small supply of food, sometimes a few lengths of timber, and a little sail cloth, the ship did not provide adequate materials for the obviously intended long stays. The construction of the shelters entailed a considerable amount of effort on the part of the sealing gang: the gathering of stones for the building of walls, the carrying of whale vertebrae and ribs, the gathering of elephant seal oil and whale bones for fuel, and the actual construction itself. This effort would not have been contemplated unless a considerable period of occupation was intended. The lack of care and investment shown by the sealing companies and ship masters in shore accommodation and conditions for its workers is on the one hand surprising, as the success of the seal hunt was dependent on the welfare of the sealers. On the other hand, it fits into a pattern of extreme parsimony exhibited by the ship owners (who often included the captain), aimed at maximising profit at whatever cost (often to the ordinary sealers): other examples are the 'lay' system of payment from which seamen often reached port in debt to the owners, and the price gouging on the sale of ship's 'slops', the clothing and comforts sold to sealers from the ship's stores at hugely inflated prices. The lay system not only disadvantaged individual sealers, but less-well-resourced ship masters, as shown in the case of Captain Christopher Burdick and the Australian Captain Richard Siddins, who both entered into agreements with Captain John Davis to combine their individual vessels with his ship and shallop in the hunt, in the 1820–21 and 1821–22 seasons, respectively. Burdick and Siddins were to receive one-third of the take of skins and Davis two-thirds, and in Siddins' case, in light of the rapidly diminishing seal numbers by 1821–22, this meant his entire voyage was a financial loss (Pearson 2018b). This unequal relationship shows the real value of owning the shallop in South Shetland sealing.

As we see in this chapter, the combination of documentary and archaeological information as independent lines of inquiry allows us to build more complete and diverse narratives about the daily life of sealer groups in South Shetlands.

FINAL THOUGHTS

As pointed out in Chapter 3, when this book was being written the *Endurance*, the lost vessel of the Antarctic explorer Sir Ernest Shackleton, crushed by sea ice and sank in 1915, was found at the bottom of the Weddell Sea. Almost immediately, the world media gave wide coverage of this finding and presented it as one of the most exciting and important discoveries in Antarctic history. The costly project, named 'Endurance22' had as its main goal the location of the vessel. Organised by the Falklands Maritime Heritage Trust, using a South African icebreaker, and supported by some prominent media companies, the project utilised the most sophisticated technological and logistical equipment.

The extent of world-wide attention raised by the discovery of the *Endurance* and the retelling of the story of the dramatic survival of Sir Ernest Shackleton and his crew gave renewed strength to the dominant narrative on Antarctica's past-hat of heroic leaders on national enterprises pitted against the sublime enmity of the Antarctic environment. In contrast to the attention afforded to these tales and objects of derring-do, the sites of the sealers in the South Shetlands, the first human interactions with the Antarctic continent, have drawn little attention from the media and minimal funding for their study. They lack the 'sexiness' of the dominant narratives, referring instead to the presence of ordinary people from subordinate groups, who went south to slaughter animals and make a profit. While sealers' sites cannot be easily attributed to specific nations, it is still doubtful whether specific nations have an interest in appropriating this past for themselves, and giving it the same prominence as other events and characters (Pearson and Salerno in press).

As we have seen in the book, although Antarctica is an international territory, the politics of the region are clearly infused with nationalistic interests and latent sovereignty claims. In this scenario, historical discourses take part of a complex political arena that influences decisions on heritage site protection and

DOI: 10.4324/9780429201257-6

management. The particular construction of dominant narratives in the region has provided a 'rather partial view of the past' (Roura 2017: 475). As we hope we have shown, archaeology is a discipline focused on the study of material culture which has the potential to shed light on different aspects of the past not always considered by official discourses.

40 years of Antarctic and sub-Antarctic archaeology

We have presented in this book an outline of the history and topics of archaeological work in Antarctica and the sub-Antarctic over a 40-year period. The archaeological resource surveys, conservation-related archaeology and the research study of the sites of the sealers and whalers show different aspects of the use of history and archaeology. This work emphasises the necessary marriage of historical and material evidence to fully understand the relationship of humans to the Antarctic environment and its resources over the past 200 years. As archaeologist Noel Broadbent has noted when thinking about the archaeology of mid-twentieth-century American bases:

> One great advantage of historic archaeology over prehistoric archaeology is that it can be matched with written sources that may reveal, among other things, the motives behind various endeavours and the details of planning and consequent successes and failures. However, written history tends also to focus on the larger picture of given events and can be biased by political, social, or other concerns of the times. Physical evidence and archaeological analysis can be used not only to test hypotheses about historical events but also, of equal importance, to provide information about the daily lives of individuals that rarely make it into historical accounts.
>
> *(Broadbent 2009: 50)*

The biases noted by Broadbent have been seen throughout the book, most notably in the emphasis on 'master narratives' such as those of the 'heroic era' exploration and 'big men' and the more recent focus on the 'science and peace' narrative of the Antarctic Treaty.

As we have seen, the history of human activity in Antarctica and the sub-Antarctic has taken place entirely in the modern era, within the framework of modernity – rising nationalism, colonialism, imperialism, mercantilism, industrialisation and globalisation. Sealing was part of a global industry based on the exploitation of marine mammals that encompassed the lands of the southern oceans (and those of the north); the nineteenth-century scientific expeditions were part of a global hunt for scientific information born of the Enlightenment, particularly that which might improve the exactness and safety of maritime navigation; whaling developed into an ocean-wide intensive industrial process, both ship and land based; and the explorations and scientific efforts from the 'heroic era' through to and after the formation of the Antarctic Treaty were part of, and responses to, international political rivalry and 'environmental imperialism'. In all of these historical phases, the study of the material evidence can provide a corrective focus on a history otherwise told only by those with a vested interest in a particular interpretation of those events and movements. It provides a means of giving voice to subaltern cultures and their stories – the ordinary seamen, sealers, whalers, expedition crewmen and gang members otherwise overlooked in the majority of histories.

In the case of the sealers, the material evidence uncovered through archaeology has helped reveal the characteristics of the sealing industry in polar regions and given voice to anonymous and forgotten actors. While the historical record recounts the putting of gangs ashore, the retrieval of skins or oil, and the overall maritime enterprise involved, the material evidence, augmented by a small handful of first-hand accounts, draws attention to the harsh working conditions imposed on sealers put ashore with insufficient resources to survive without extensive utilisation of local food and material resources. Considering the material evidence and placing it in its cultural landscape setting also illuminates the physical conditions that some sealers found liberating after the tight controls and regulated labour experienced aboard ship, a human response expressed in just one or two reminiscences. The lived experiences so manifestly illustrated by the archaeological work offer a more humanised counter-narrative to the story of global exploitative economics in an industrialised world that still tends to dominate the histories of sealing and whaling. The material evidence weaves a picture of self-reliance and adaptation to new circumstances: the use of local resources of rock, whale bones and seal skins for shelter; the eating of seal and penguin organs and meat to augment the limited food carried ashore; the repair or manufacture of clothing and especially footwear with seal hide; the creation of games and smoking pipes to pass the time; and the organisation of work and life within very restricted sheltered spaces. This evidence and the limited documentary record can be combined to try to reconstruct the social practices and interpersonal relations that accompanied and enabled these adaptive responses.

Looking at the picture of sealing on the sub-Antarctic islands and the South Shetland Islands, an important distinction can be seen between the ideals of a capitalist commercial operation and the reality of its as-lived implementation on the ground. The economic historical view would see the organisation of production through standardised processes, division of labour, regulation of time and hierarchical power structures as the tools underpinning the global success of an industry. Once off-ship, any chance of implementing that sort of regulated system in the Antarctic sealing industry evaporated. Small under-resourced gangs set ashore for unpredictable periods of time to gather an uncertain amount of the seal resource in a harsh environment had other priorities – the first being to survive. The structure of hierarchy able to be maintained on-board a ship was difficult if not impossible for one officer to impose without back-up on the rest of a gang of sealers once ashore. The gathering of seal products became just one consideration shared with making shelter, gathering and preparing food, and finding ways of replicating the resources available aboard but not brought ashore. To a probably large degree, the balance of power in the social hierarchy shifted, and documentary evidence indicates that it shifted more substantially as time ashore lengthened, especially if the seal resource diminished because of over-exploitation or seasonal absences of seals.

The archaeology of the 'heroic era' huts also throws some light on the seldom documented activities and responses of the members of the support parties. The translation of paper-based plans into physical reality, both in terms of the buildings and the base layout and use, by men not necessarily skilled or experienced in such work, is evidenced at several sites where the material evidence shows poor workmanship and the integration of features that not only varied from the carefully prepared plans but also might have actually influenced the validity of the scientific observation made (such as iron fixings in huts built for magnetic observations). The distribution of waste refuse and the construction of makeshift annexes for work and storage around the huts cast light on the human responses to the harsh winter environment. This corpus of physical evidence, added to the growing list of uncensored personal diaries and papers of ordinary party members being published, begins to produce a historical perspective from the viewpoint different from that of their leaders. Other aspects of the archaeological work at the 'heroic era' huts have been used, intentionally or unintentionally, to give material reinforcement of the transmission of the dominant narratives of their purpose and use. The archaeology at the Ross Sea huts, for example, has been used in part in the recreation of the idealised tableau of 'original' occupation of the huts: the 'archaeology of nostalgia' as Ximena Senatore has called it (Senatore 2019). The needs of conservation programs have determined the nature of archaeological work at the 'heroic era' huts in the past. There has been little space or resources available for broader research questions, although some have attempted such work (e.g. on Mawson's Huts artefacts scatter).

Archaeology on the sub-Antarctic islands has largely been initiated as a part of surveys to define the resources of the islands for management purposes. This has led to systematic cataloguing of sites as a basis for protection and monitoring, and triggered more intensive archaeological work at a number of sites. Such broad-based survey work has been rare in the Antarctic itself, where management responsibility is much more narrowly defined, usually limited to the boundaries of a scientific station. In part as a result of this, there are no systematically created inventories of cultural heritage sites for the continent or regions, with the exception of the South Shetland Islands where the survey of islands for sealing and other sites has been ongoing for several decades but is still not complete.

The study of the material and documentary evidence from the perspective of subaltern cultural experience has led a number of the archaeologists to look at other ways to respond to the lived experience of sealers – how they might have related to the Antarctic landscape and how that experience might be understood by modern observers (Salerno et al. 2010; Salerno and Zarankin 2014). The concepts of how sealers might have perceived their environment through sensoriality, movement and emotion have been explored, and ways of presenting the sealing experience to domestic audiences through scanning and 3D technologies, reproduction of artefacts and their presentation in simulated environments have been developed.

The study of the archaeological sites in the Antarctic and sub-Antarctic has stressed the importance of understanding the interconnection of the exploration, exploitation and scientific study of the southern oceans with the other interrelated parts of the world. The roles of home ports, supply ports, other areas exploited by sealers and whalers, sources of labour, markets and end-users, scientific institutions, ships as floating worlds and many other factors are all intertwined in the production of what appears on the ground. To understand a global industry like sealing, say, and the sites it produces, the range of these global interconnections has to be identified and understood.

Ways forward

We noted in Chapter 1 some of the characteristics of research and conservation in Antarctica, and in particular the great logistical costs of archaeological projects. A future development that may relieve this logistical problem is the potential use of 'volunteer science' to carry out professionally guided survey, research and conservation works. This approach has been used elsewhere in the world in various forms over the last few decades. Such an approach could be based on tourist vessels and air access, with volunteers contributing to the cost of the work over and above their own travel costs. It could add new opportunities to the mix of offerings of the tourism industry, widen the public understanding of Antarctica's past and present and help fund ongoing work. Tentative attempts at this approach have occurred in the past, with some research teams gaining access to sites via cruise ships, but for it to work more effectively, the nature and duration of the tourist presence would have to expand.

We have also drawn attention to the ways the material evidence can be used to 'democratise' the stories of Antarctica for a broader public audience. Stimulating public interest in archaeological work is an important tool in influencing political decisions about funding, and also, related the previous paragraph, the enticement of tourists to Antarctica for research-related visits. The continued development of digital technologies for recording and reproducing archaeological sites and objects is both anticipated and welcomed as ways of making information about archaeological sites more accessible and finding new ways of engaging the public with them.

There are many remaining gaps in archaeological research: the pursuit of historical and 'anthropological' research questions (as distinct from resource management or conservation questions) at 'heroic era' huts, scientific bases, whaling stations and on the sub-Antarctic islands; and the completion of the survey and study of the sealing sites in the South Shetlands, including the tantalising nexus of shipwreck sites and survivor camps known to exist there.

Challenges remain in the governance of cultural heritage sites and archaeology in the Antarctic, and additional support would be useful in the sub-Antarctic. As has been indicated, the absence of a 'national government' for Antarctica has meant that the undertaking of such things as wide-spread heritage site surveys and inventories has not occurred. Even in the South Shetlands where intense study has been underway for decades, the systematic survey of all the islands has not yet been completed, primarily because of logistical (mainly transport) constraints. Interested bodies have developed tools to try to help in setting standards for archaeological work in the Antarctic. For example, the *Antarctic Archaeology Guidelines* developed by the International Polar Heritage Committee of ICOMOS (IPHC) have been adopted to apply standards that exist internationally to be tailored for Antarctic conditions. The IPHC also provides independent expert comment on issues of site identification, conservation and management as they arise.

We hope this book has into perspective the role of archaeology in the study of the past of Antarctica and the sub-Antarctic over the past 40 years. The different uses of archaeology have been discussed, as has the ongoing potential of the archaeological study of the material remains of human activities to add substantially to the understanding of our presence on the continent and the surrounding islands. Its power to give previously unrepresented groups visibility provides a counter-balance to the master narratives that have dominated Antarctic historiography over the last century or more, and hopefully reduces their dominance. Each kaolin pipe, shoe, bone or any other recovered fragment contains the germ of a new story to be told, and from these narratives it is hoped that a better understanding of humankind's interaction with Antarctica and the sub-Antarctic can arise, forming a better basis for our future interactions with the region.

BIBLIOGRAPHY

AAP Mawson's Huts Foundation. 1997. *Initial Environmental Evaluation: AAP Mawson's Huts Foundation Conservation Program 1997–98*. Unpublished report for the AAP Mawson's Huts Foundation.

AAP Mawson's Huts Foundation. 2001. *Report of the AAP Mawson's Huts Foundation Expedition 2000–2001*. Unpublished report for the AAP Mawson's Huts Foundation.

Akkerman, A. 2016. *Phenomenology of the Winter-City: Myth in the Rise and Decline of Built Environments*. Springer, New York.

Aldazabal, V. 2007. Recuperación y puesta en Valor de la casa Omond. Orcadas del Sur. In Vairo, C., Capdevila, R., Aldazabal, V. and Pereira, P. (eds.), *Antártida, Patrimonio Cultural de la Argentina. Museos, Sitios y Refugios Históricos de la Argentina*. Zaguier and Urruty, Ushuaia, pp. 43–61.

Aldazabal, V. 2012. Hilvanando historias. Textiles recuperados en Omond House, Islas Orcadas del Sur. *Revista Estudios Hemisféricos y Polares*, 13(3): 1–17.

Aldazabal, V. 2017. El cuero en la vida cotidiana: registros de la Base Orcadas – 1904. *Revista de Arqueología Histórica Argentina y Latinoamericana*, 11(1): 43–62.

Alkmim, G.R. de. 2015. *Os Sapatos dos Lobeiros-Baleeiros: Práticas de Calçar do Século XIX nas Ilhas Shetland do Sul (Antártica)*. Master's diss. Universidade Federal de Minas Gerais, Belo Horizonte.

Almevik, G., Avango, D., Contissa, V., Fontana, P., Lindström, K. and Westin, J. 2021. *Built Cultural Heritage in Antarctica: Remains and Uses of the First Swedish South Polar Expedition 1901–1903*. Swedish National Heritage Board, Visby.

Ames, N. 1830. *A Mariner's Sketches, Originally Published in the Manufacturers and Farmers Journal, Providence*. Cory, Marshall and Hammond, Providence.

Amorim, C. 2019. *Cachimbos de Caulim em Terras Geladas: Sobre Foqueiros e Baleeiros mediante seu Fumar na Antártica*. BA diss. Universidade Federal de Minas Gerais, Belo Horizonte.

Anderson, A. 2005. Subpolar settlement in South Polynesia. *Antiquity*, 79: 791–800.

Anderson, A. and O'Regan, G. 2000. To the final shore: Prehistoric colonisation of the subantarctic islands in South Polynesia. In Anderson, A.J. and Murray, T. (eds.),

Australian Archaeologist. Collected Papers in Honour of Jim Allen. Coombs Academic Publishing, The Australian National University, Canberra, pp. 440–454.

Aragón, M. 1991. La última singladura del navío San Telmo. *Boletín Antártico Chileno*, 10: 38–40.

ATCM. Papers of the Antarctic Treaty Consultative Meetings, accessible by meeting number or date at the Antarctic Treaty Data Base. Retrieved from https://www.ats.aq/devAS/ToolsAndResources/AntarcticTreatyDatabase?lang=e

Australian Antarctic Division. 2013. *Mawson's Huts Historic Site: Management Plan 2013–2018*. Australian Antarctic Division, Kingston.

Bargagli, R. 2005. *Antarctic Ecosystems: Environmental Contamination, Climate Change, and Human Impact*. Springer, Heidelberg.

Barr, S. 2018. Twenty years of protection of historical values in Antarctica under the Madrid Protocol. *The Polar Journal*, 8(2): 241–264.

Barr, S. and Pearson, M. 2021. The Polar Regions. In Goetcheus, C.L. and Brown, S. (eds.), *Routledge Handbook of Cultural Landscapes*. Routledge, London.

Basberg, B.L. 2004. *The Shore Whaling Stations at South Georgia: A Study in Antarctic Industrial Archaeology*. Novus Forlag, Oslo.

Basberg, B.L. 2007. History of whaling. In Riffenburgh, B. (ed.), *Encyclopedia of the Antarctic*. London, Routledge, pp. 1072–1076.

Basberg, B.L., Burton, R. and Headland, R. 2018. Fur seal imports to Britain, 1800–1840: New estimates in context. In Headland, R.K. (ed.), *Historical Antarctic Sealing Industry. Proceedings of an International Conference in Cambridge 16–21 September 2016*. Scott Polar Research Institute, Occasional Publication Series, Cambridge, pp. 79–86.

Basberg, B.L. and Headland, R.K. 2008. *The 19th Century Antarctic Sealing Industry: Source, Data and Economic Significance*. Institutt for Samfunnøkonomi, Bergen.

Basberg, B.L., Nævestad, D. and Rossnes, G. 1996. Industrial archaeology at South Georgia: Methods and results. *Polar Record*, 32(180): 51–66.

Berguño, J.B. 1993a. Las Shetland del Sur: El ciclo lobero. Primera parte. *Boletín Antártico Chileno*, 12(1): 5–13.

Berguño, J.B. 1993b. Las Shetland del Sur: El ciclo lobero. Segunda parte. *Boletín Antártico Chileno*, 12(2): 2–9.

Bertrand, K.J. 1971. *Americans in Antarctica, 1775–1948*. American Geographical Society, New York.

Besse, J.M. 2006. Las cinco puertas del paisaje; ensayo de una cartografía de las problemáticas paisajeras contemporáneas. In Maderuelo, J. (ed.), *Paisaje y Pensamiento*. Adaba, Madrid, pp. 145–171.

Bezerra, M. 2013. Os sentidos contemporâneos das coisas do passado: Reflexões a partir da Amazônia. *Revista de Arqueologia Pública*, 7(2): 107–122.

Bloom, L. 1993. *Gender on Ice: American Ideologies of Polar Expeditions*. Minnesota University Press, Minnesota.

Bonner, W.N. 1985. Impact of fur seals on the terrestrial environment at South Georgia. In Siegfried, W.R., Condy, P.R. and Laws, R.M. (eds.), *Antarctic Nutrient Cycles and Food Webs*. Springer, Berlin, pp. 641–646.

Borrero, L.A. 1999. The prehistoric exploration and colonisation of Fuego-Patagonia. *Journal of World Prehistory*, 13(3): 321–355.

Boshoff, J.J., Haart, D. and Loock, J. 1997. *Survey of Historical Sites on Marion Island*. Report for the Prince Edward Islands Management Committee, by Representatives from the National Monuments Council and the Cultural History Museum Maritime Division. Retrieved from https://alp.lib.sun.ac.za/bitstream/handle/123456789/2426/marion_island_survey_historical_sites.pdf?sequence=4&isAllowed=y

Boshoff, J., Niekerk, T. van and Wares, H. 2015. Preliminary investigations on the wreck of the SS *Solglimt*, Marion Island. *Bulletin of the Australasian Institute of Maritime Archaeology*, 39: 53–59.

Boshoff, J.J. and Schalkwyk, J.A. van. 1999. *The Archaeology of Sealing Sites on Marion Island*. Unpublished report for the South African Heritage Resources Agency.

Brabo, A. 2019. *Técnicas de Modelagem Tridimensional Aplicadas à ConservaçÃo de Acervos Arqueológicos: Documentação de Cachimbos do Acervo do LEACH-UFMG*. BA diss. Universidade Federal de Minas Gerais, Belo Horizonte.

Broadbent, N.D. 1994. An archaeological survey of Marble Point Antarctica. *Antarctic Journal of the United States*, 29(4): 3–6.

Broadbent, N.D. 2009. From ballooning in the Arctic to 10,000-foot runways in Antarctica: Lessons from historical archaeology. In Krupnik, I., Land, M.A. and Miller, S.E. (eds.), *Smithsonian at the Poles: Contributions to International Polar Year Science*. Smithsonian Institution Scholarly Press, Washington, D.C., pp. 49–60.

Broadbent, N.D., Blee, C.H. and Spude, R.L. 1992. Reclaiming Antarctic history. The restoration of the East Base, Antarctica. *Antarctic Journal of the United States*, 27(2): 14–27.

Broadbent, N.D. and Rose, L. 2002. Historical archaeology and the Byrd legacy: The United States Antarctic Service Expedition, 1939–41. *The Virginia Magazine of History and Biography*, 110(2): 237–258.

Burdick, C. 1821. Log of the ship '*Huntress*' 1820–21. Original manuscript in collection of E.A. Stackpole, copy used in SPRI MS 440.MJ.

Burkitt, D.M. 1996. *Report on the 1996 Work Programme at Port Lockroy*. British Antarctic Survey.

Burnett, D.G. 2012. *The Sounding of the Whale: Science and Cetaceans in the Twentieth Century*. University of Chicago Press, Chicago.

Burns, R. 2007. Women in Antarctica: From companions to professionals. In Riffenburgh, B. (ed.), *Encyclopedia of the Antarctic*. Routledge, London, pp. 1092–1096.

Burton, H.R. and Williams, D.L. 1985. *Heard Island ANARE 1985 Report*. Australian Antarctic Division.

Burton, R. 2018. From shoes to shawls: Utilization of 'South Seas' fur seal pelts in late 18th and early 19th Century England'. In Headland, R.K. (ed.), *Historical Antarctic Sealing Industry: Proceedings of an International Conference in Cambridge 16–21 September 2016*. Scott Polar Research Institute, Cambridge. Occasional Publication Series, pp. 87–93.

Busch, B. 1980a. Elephants and whales: New London and Desolation, 1840–1900. *The American Neptune*, 40(2): 117–126.

Busch, B. 1980b. *Master of Desolation: The reminiscences of Capt. Joseph J. Fuller*. Mystic Seaport Museum Inc., Mystic.

Busch, B. 1985. *The War against the Seals: A History of the North American Seal Fishery*. McGill-Queen's University Press, Kingston.

Butler, J. 2000. Corpos que pesam: Sobre os limites discursivos do "sexo". In Louro, G. (ed.), *O corpo Educado: Pedagogias da Sexualidade*. Autêntica, Belo Horizonte, pp. 151–172.

Butler, J. 2003. *Problemas de Gênero: Feminismo e Subversão da Identidade*. Civilização Brasileira, Rio de Janeiro.

Cabrera, L. 2002. Arqueología histórica en el continente Antártico; tras el fantasma del San Telmo. *Actas del I Congreso de Arqueología Histórica*, Corregidor, Buenos Aires, pp. 939–945.

Calder, J. 1984. *The Circumnavigation of Van Diemen's Land in 1815 by James Kelly and in 1824 by James Hobbs, Edited from their own Accounts by James Calder*. Sullivans Cove, Hobart.

Campbell, R.J. 2000. *The Discovery of the South Shetland Islands: The Voyage of the Brig Williams 1819–1820 As Recorded in Contemporary Documents and the Journal of Midshipman C.W. Poynter.* Hakluyt Society, London.

Capdevila, R. n.d. Argentina and the historic shelters of the Swedish Expedition 1901–1903. *Museo Marítimo de Usuaia.* Retrieved from https://museomaritimo.com/en/historic-shelters

Capdevila, R.Y. 1990. Arqueología histórica en la Antártida. *Histarmar. Historia y Arqueologia Marítima.* Retrieved from http://www.histarmar.com.ar/Antartida/ArqueologiaHist-1.htm

Capdevila, R.Y. and Ageitos, J.M. 1993. Restauración de la choza sueca de Bahía Esperanza. *Contribuciones del Instituto Antártico Argentino*, 417: 1–20.

Capdevila, R.Y. Ageitos, J.M. and Melemenis, V. 1994. Restauración de la choza sueca de Bahía Esperanza—II. *Contribuciones del Instituto Antártico Argentino*, 423: 1–11.

Capdevila, R.Y. Ageitos, J.M., Orsetti, A., Pereira, A., Melemenis, V. and Fourcade, D. 2000. La cabaña sueca de la Isla Cerro Nevado. *Contribuciones del Instituto Antártico Argentino*, 451: 1–14.

Capdevila, R.Y. and Commerci, S.M. 1986. *Historia Antártida Argentina.* Dirección Nacional del Antártico, Instituto Antártico Argentino, Buenos Aires.

Cartajena, I. and Labarca, R. 2007. Patrones históricos de cuereo en los marinos: La evidencia de los Abrigos 1 y 2 de Playa Yámana (Isla Livingston, Antártica chilena). *Magallania*, 35(1): 95–105.

Carter, P.A. 1979. *Little America - Town at the End of the World.* Columbia University Press, New York.

Caviglia, S. 2015. *Malvinas: Soberanía, Memoria y Justicia. Vol. II: Balleneros, Loberos, Misioneros, S. XVIII-XIX.* Ministerio de Educación de la Provincia de Chubut, Rawson.

Cazenave de la Roche, A. "Pesquisas sobre el descubrimiento de la Antártida: tras las estelas del *San Telmo* y del *Williams*." In *Magallánica. Revista de Historia Moderna*, vol. 6, no11 (2019). Dossier: "Itinerarios de Globalización: Historia y Arqueología Marítima entre España y América", coordinated by Ana Crespo Solana. 2019, pp. 276-317.

Chapman, A. 2010. *European Encounters with the Yamana People of Cape Horn, before and after Darwin.* Cambridge University Press, Cambridge.

Chapman, T. 1810. *A Narrative of the Case of Mr. Thomas Chapman who First Discovered the Means of Making the Fur of the Seal Available to the Manufacturers of Hats, Shawls, & Superfine Cloths: With the Evidence Adduced before a Committee of the Honorable House of Commons.* Kemmish and Son, London.

Chipman, E. 1986. *Women on the Ice.* Melbourne University Press, Melbourne.

Clark, A.H. 1887. The Antarctic fur-seal and sea-elephant industry. In Goode, G.B. (ed.), *The Fisheries and Fishery Industries of the United States.* Vol II. Government Printing Office, Washington, pp. 400–467.

Clark, G. and de Biran, A. 2010. Geophysical and archaeological investigation of the survivor-camp of the *Antelope* (1783) in the Palau Islands, Western Pacific. *The International Journal of Nautical Archaeology*, 39(2): 345–356.

Clark, L. and Wishart, E. 1990. *Historical documentation of Wilkes Station, Antarctica 1988–89.* Report for the Australian Antarctic Division.

Clark, L. and Wishart, E. 1991. *Historical documentation of Old Casey Station, Antarctica.* Queen Victoria Museum and Art Gallery, Launceston.

Clark, L. and Wishart, E. 1993. *Heritage Assessment of Mawson Station, Antarctica.* Report for Queen Victoria Museum and Art Gallery.

Clarke, C.M. 1854. *Glances at Life upon the Sea, or Journal of a Voyage to the Antarctic Ocean, in the Brig Parana, of Sag Harbor, L. I., in the Years '53 '54; Description of Sea-Elephant*

Hunting among the Icy Islands of South Shetland, Capture of Whales, Scenery in the Polar Regions, &c. Charles H. Pelton, Middleton.

Clarke, W.B. 1850. *Narrative of the Wreck of the "Favourite" on the Island of Desolation: Detailing the Adventures, Sufferings, and Privations of John Nunn, an Historical Account of the Island, and its Whale and Seal Fisheries.* William Edward Painter, London.

Cochran, C.C. (ed.). 1990. *Heritage Management Plan for the Historic Sites of the Ross Sea Region.* Antarctic Heritage Trust, Christchurch.

Cochran, C.C. 1996. *Report on the Conservation of Port Lockroy Base A, Goudier Island, Antarctic Peninsula.* Report for the British Antarctic Survey.

Collis, C. 2009. The Australian Antarctic Territory: A man's world? *Signs: Journal of Women in Culture and Society*, 34(3): 514–519.

Comerci, S.M. 1981. Arqueología histórica en la Antártida: Los trabajos de la República Argentina en la Isla Cerro Nevado durante la campaña antártica de verano 1979–80'. *Karukinka, Cuaderno Fueguino*, 26: 180–188.

Comerci, S.M. 1983. Arqueología antártica: Los trabajos de la República Argentina en la Isla Cerro Nevado durante las campañas antárticas 1979–80 y 1980–81. *Contribuciones del Instituto Antártico Argentino*, 291: 1–25.

Commonwealth of Australia. 2014. *Heard Island and McDonald Islands Marine Reserve Management Plan – 2014–2024.* Australian Antarctic Division, Department of the Environment.

Constantinescu, F.C. and Torres, D.N. 1995. Análisis bioantropológico de un cráneo humano hallado en Cabo Shirreff, Isla Livingston, Antártica. *Boletín del Instituto Antártico Chileno*, serie científica, 45: 89–99.

Cook, N.D. 2004. *Demographic collapse: Indian Peru, 1520–1620.* Cambridge University Press, Cambridge.

Cooper, J.F. 1849. *The Sea Lions, or The Lost Sealers.* Stringer & Townsend, New York.

Cooper, J. and Avery, G. (eds.). 1986. *Historic Sites at the Prince Edward Islands.* South African National Scientific Programmes Report No. 128. Foundation for Research Development, Council for Scientific and Industrial Research, Pretoria.

Cooper, J., Boshoff, J., Harboe-Ree, C. and van Niekerk, T. 2018. The wreck of the Norwegian sealer *Solglimt* on sub-Antarctic Marion Island: A combined archaeological and historical investigation. In Headland, R.K. (ed.), *Historical Antarctic Sealing Industry. Proceedings of an International Conference in Cambridge 16–21 September 2016.* Scott Polar Research Institute, Occasional Publication Series, Cambridge, pp. 95–100.

Cordeiro, R. 2019. Tracing the birth of a global economy. *Penguin News*, April 5: 12.

Connell, R. 2005. *Masculinities.* University of California Press, Berkeley.

Courbon, P. 2012. *Le Chaudron de Crozet.* Report in TAAF website. Retrieved from http://www.chroniques-souterraines.fr/dossiers/Archeologie/00_Dossiers/crozet.pdf

Courbon, P. and Le Mouël, J-F. 2009. Mission archéologique internationale en Baie de l'Observatoire – Illes Kerguelen, 2006–2007. *ArchaeObs, Terres Australes et Antarctiques Francaises*, May. Retrieved from http://www.chroniques-souterraines.fr/dossiers/view/archeobs.html (accessed 10/6/19)

Creighton, M. 1995. *Rites and Passages: The Experience of American Whaling, 1830–1870.* Cambridge University Press, Cambridge.

Cruz, M.J. 2011. *Aproximaciones a las Prácticas Alimenticias de los Grupos Foqueros (Islas Shetland Del Sur, Siglo XIX).* BA diss. Universidad de Buenos Aires, Buenos Aires.

Cruz, M.J. 2014. *Incorporando Comidas e Contextos: A Alimentação e o Corpo nos Grupos Foqueiros nas Setland do Sul (Antártica, Século XIX).* Master's diss. Universidade Federal de Minas Gerais, Belo Horizonte.

Cruz, M.J. 2016. Incorporando comidas e contextos: A alimentação dos grupos foqueiros nas Setland do Sul (Antártica, século XIX). In Soares, F.C. (ed.), *Comida, Cultura e Sociedade. Arqueologias da Alimentação no Mundo Moderno*. Estudos Contemporâneos na Arqueologis 2. Editora Universitaria UFPE, Recife, pp. 169–190.

Cruz, M.J. 2017. La alimentación sentida. Cocinando algunas reflexiones. In Pellini, J., Zaranki, A. and Salerno, M. (eds.), *Sentidos Indisciplinados. Arqueología, Sensorialidad y Narrativas Alternativas*. JAS, Madrid, pp. 113–156.

Cruz, M.J. 2018. Food and feeding of sealers on Livingston Island, South Shetland Islands. In Headland, R.K. (ed.), *Historical Antarctic Sealing Industry. Proceedings of an International Conference in Cambridge 16–21 September 2016*. Scott Polar Research Institute, Occasional Publication Series, Cambridge, pp. 101–106.

Cruz, M.J. 2019. *Memórias de um Mundo Congelado. A Indústria Lobeira e as Experiências Antárticas no Século XIX. PhD diss.* Universidade Federal de Minas Gerais, Belo Horizonte.

Csordas, T. 1990. Embodiment as a paradigm for anthropology. *Ethos*, 18(1): 5–47.

Cumpston, J.S. 1968. *Macquarie Island*. Antarctic Division, Department of External Affairs, Melbourne.

Cumpston, J.S. 1973. *First Visitors to Bass Strait*. Roebuck Books, Canberra.

Currie, S. 2001. *Thar She Blows: American Whaling in the Nineteenth Century*. Lerner Publications, Minneapolis.

Darlington, J. 1957. *My Antarctic Honeymoon*. Frederick Muller, London.

Davies, M. 1982. *Macquarie Island: Archaeological Report*. Unpublished report for the National Parks and Wildlife Service of Tasmania.

Davis, G.A. 2017. A history of McMurdo Station through its architecture. *Polar Record*, 53(2): 167–185.

Davis, J. 1822. Log of the New Haven sealer '*Huron*' 1821–22, kept by John Davis. Original manuscript in Yale University Library, copy used at SPRI MS 442.

Davis, W. 1874. *Nimrod of the Sea; or the American Whalemen*. Harper & Brothers, New York.

Debenham, F. (ed.). 1945. *The Voyage of Captain Bellingshausen to the Antarctic Seas 1819–1821*. Hakluyt Society, London.

Decker, R.O. 1973. *Whaling Industry of New London*. G. Schumway, York.

Deetz, J. 1977. *In Small Things Forgotten: The Archaeology of the Early American Life*. Anchor Press, New York.

Delano, A. 1817. *A Narrative of Voyages and Travels in the Northern and Southern Hemispheres; Comprising Three Voyages Round the World; Together with a Voyage of Survey and Discovery, in the Pacific Ocean and Oriental Islands*. E.G. House, Boston.

Denham, T. 2017. Landscape archaeology. In Gilbert, A.S. (ed.), *Encyclopedia of Geoarchaeology*. Springer, Dordrecht, pp. 464–468.

Dingwall, P.R. and Jones, K. 2009. The Enderby settlement (1849–52): Archaeological reconstruction of a British colonial settlement at the Auckland islands. In Dingwall, P.R, Jones, K.L. and Egerton, R. (eds.), *In Care of the Southern Ocean: An Archaeological and Historical Survey of the Auckland Islands*. New Zealand Archaeological Association Monograph, Auckland, pp. 59–85.

Dodds, K. 2009. Settling and unsettling Antarctica. *Signs: Journal of Women in Culture and Society*, 34(3): 505–509.

Dodds, K. 2021. The Antarctic Treaty, and associated charts and maps. In Foscari, G. (ed.), *Antarctic Resolution*. Lars Müller Publishers, Zurich, pp. 114–123.

Donaghue, S., Jacka, T.H., Morgan, V. and Lazer, E. 2014. The microclimate of Mawson's Hut based on snow and ice core analysis. *Polar Record*, 50(1): 37–42.

Dowdeswell, J.A., Batchelor, C.L., Dorschel, B., Benham, T.J., Christie, F.D.W., Dowdeswell, E.K., Montelli, A., Arndt, J.E. and Gebhardt, C. 2020. Sea-floor and sea-ice conditions in the western Weddell Sea, Antarctica, around the wreck of Sir Ernest Shackleton's Endurance. *Antarctic Science* 32(4): 1–13.

Downes, E.M. 2010. *Barrels on a Beach: Power, Authority and the Crew of the Trinity at Heard Island 1880–1882*. Master's diss. University of Melbourne, Melbourne.

Downes, M. 1996. Indexing sealer's logbooks from Heard Island. *ANARE Research Notes*, 97: 1–96.

Downes, M. and Downes, E. 2006. Sealing at Heard Island in the nineteenth century. In Green, K. and Woehler, E.J. (eds.), *Heard Island: Southern Sentinel*. Surry Beatty & Sons, Chipping Norton, pp. 184–195.

Dürbeck, G., Schaumann, C. and Sullivan, H. 2015. Human and non-human agencies in the Anthropocene. *Ecozon@*, 6(1): 118–136.

Egerton, R., Burgess, S., Petchey, P. and Dingwall, P.R. 2009. The Auckland Islands shipwreck era. In Dingwall, P.R., Jones, K.L. and Egerton, R. (eds.), *In Care of the Southern Ocean: An Archaeological and Historical Survey of the Auckland Islands*. New Zealand Archaeological Association Monograph, Auckland, pp. 123–190.

Eights, J. 1838. Description of the New South Shetland Isles. In Fanning, E. (ed.), *Voyages to the South Seas, Indian and Pacific Oceans, China Sea, Northwest Coast, Feejee Islands, South Shetlands, etc.* William H. Vermilye, New York, pp. 198–199.

Ellis, R. 1991. *Men and Whales*. Knopf, New York.

Esteves, J., Piana, E., Schiavini, A. and Juan-Muns, N. 2001. Archaeological analysis of shell middens in the Beagle Channel, Tierra del Fuego Island. *International Journal of Osteoarchaeology*, 11(1–2): 24–33.

Evans, S-L. 2011. Icy heritage – managing historic sites in the Antarctic: Pristine wilderness, anthropogenic degradation or cultural landscape? *The Polar Journal*, 1(1): 87–100.

Fanning, E. 1833. *Voyages Round the World; with Selected Sketches of Voyages to the South Seas, North and South Pacific Oceans, China etc. between 1792 and 1832*. Collins and Hannay, New York.

Fastier, A., Clendon, R., Rogan, D. and Ritchie, N. 2006. K440 Field Report Summer Season 2005 for the Historic Huts on Ross Island, Antarctica. Archaeologist's section, pp. 15–23.

Fildes, R. 1820–21. *A journal of a Voyage Kept on Board Brig Cora of Liverpool bound to New South Shetland*. Public Records Office Series: Adm. 55, Admiralty and Secretariat. Log books etc. Supplementary, Series II, explorations. PRO Australian Joint Copying Project (AJCP) reel 1599, piece 143, pp. 1–41.

Fildes, R. 1821a. *Remarks made during a Voyage to New South Shetland*. Public Records Office Series: Adm. 55, Admiralty and Secretariat. Log books etc. Supplementary, Series II, explorations. PRO Australian Joint Copying Project (AJCP) reel 1599, piece 143, pp. 45–66.

Fildes, R. 1821b. *A Journal of a Voyage from Liverpool towards New South Shetland on a Sealing and Sea Elephant Adventure Kept on Board Brig Robert of Liverpool, Robt. Fildes*. Public Records Office Series: Adm. 55, Admiralty and Secretariat. Log books etc. Supplementary, Series II, explorations. PRO Australian Joint Copying Project (AJCP) reel 1599, piece 143, pp. 67–96.

Fitte, E. 1962. *El Descubrimiento de la Antártida. Crónica de los Homres y Barcos que Exploraron las Aguas de las Shetland del Sur*. EMECÉ, Bueos Aires.

Fogg, G.E. 1992. *A History of Antarctic Science*. Cambridge University Press, Cambridge.

Foscari, G. 2021a. (ed.). *Antarctic Resolution*. Lars Müller Publishers, Zurich.

Foscari, G. 2021b. Domesticity in the extreme. In Foscari, G. (ed.), *Antarctic Resolution*. Lars Müller Publishers, Zurich, p. 448.

Funari, P. 1999. Etnicidad, identidad y cultura material: Un estudio del cimarrón Palmares, Brasil, siglo XVII. In Zarankin, A. and Acuto, F. (eds.), *Sed Non Satiata: Teoría Social en la Arqueología Latinoamericana Contemporánea*. Del Tridente, Buenos Aires, pp. 77–96.

Funari, P., Jones, S. and Hall, M. 1999. Introduction: Archaeology in history. In Funari, P., Hall, M. and Jones, S. (eds.), *Historical Archaeology. Back from the Edge*. Routledge, London, pp. 1–20.

Fyfe, R. 1990. *Ross Island Historic Huts: Cape Evans, Cape Royds, Hut Point 1989/90: Report on Archaeological Field Work and Future Management Considerations*. Report to the Antarctic Heritage Trust.

Fyfe, R. 1992. *Ross Island Historic Sites: 1991–92*: Report to the Antarctic Heritage Trust.

Galloway, P. 2006. Material culture and text: Exploring the spaces within and between. In Hall, M. and Silliman, S. (eds.), *Historical Archaeology*. Blackwell, Oxford, pp. 42–64.

Gibb, R. and McCurdy, D. 2013. The South Georgia Island shore-based whaling stations survey project. *Archaeology in New Zealand*, 56(2): 40–52.

Gibb, R., McCurdy, D. and Farrell, R. 2011. The use of laser scanning to record historic huts in Antarctica. *Archaeology in New Zealand*, 54(2): 131–143.

Gibbs, M. 2002. *Maritime Archaeology and Behaviour during Crisis: The Wreck of the VOC Cultural Change*. Routledge, New York, pp. 66–86.

Gibbs, M. 2003. The archaeology of crisis: Shipwreck survivor camps in Australasia. *Historical Archaeology*, 37(1): 128–145.

Gibbs, M. 2006a. Cultural site formation processes in maritime archaeology: Disaster response, salvage and Muckelroy 30 years on. *The International Journal of Nautical Archaeology*, 35(4): 4–19.

Gibbs, M. 2006b. Maritime archaeology at the land–sea interface. In Staniforth, M. and Nash, M. (eds.), *Maritime Archaeology: Australian Approaches*. Springer, New York, pp. 69–81.

Gillespie, A. 2005. *Whaling Diplomacy. Defining Issues in International. Environmental Law*. Edward Elgar Publishing, Cheltenham.

Godfrey, I. 2002. *Mawson's Huts 2002 Field Work Report*. Unpublished report for the Australian Antarctic Division.

Goldberg, F., Wiklander, L. and Capdevila, R. 2001. *The Swedish Hut in Antarctica: The Construction and Preservation of the Nordenskjöld Hut on Snow Hill Island, Antarctica*. The Stockholm Building Society, Lidingö.

González-Ruibal, A. 2007. Arqueología simétrica: Un giro teórico sin revolución paradigmática. *Complutum*, 18: 283–285.

Goode, G.B. (ed.) 1887. *The Fisheries and Fishery Industries of the United States*. Government Printing Office, Washington.

Goodridge, C.M. 1832 (and editions 1839, 1841). *Narrative of a Voyage to the South Seas and the Shipwreck of the Princess of Wales Cutter, on One of the Crozets, Uninhabited Islands; with an Account of a Two Years' Residence on them by the Crew*. Hamilton and Adams, London.

Gould, R.T. 1941. The charting of the South Shetlands, 1819–28. *The Mariner's Mirror*, 27(3): 206–242.

Graham, T.A. 1989. *Cultural Resource Management of the Prince Edward Islands*. BA diss. University of Cape Town, Cape Town.

Grant, A. 1998. *Report by Expedition Leader, Alan Grant, AAP Mawson's Huts Expedition 1997–98 – Filed Leader's Report, Condensed Version*. Unpublished Report for the Australian Antarctic Division and AAP Mawson's Huts Foundation.

Grinbergs, A.M. 2000. *ANARE Station Atlas Cove, Heard Island: Condition Assessment and Preliminary Notes Relating to the Conservation and Management of Cultural Heritage Values*. Unpublished report to the Australian Antarctic Division.

Hacquebord, L. 1992. Hector Station on Deception Island (South Shetland Islands, Antarctica): An environmental assessment study of a whaling station. *Circumpolar Journal*, 7(1–2): 72–97.

Hacquebord, L. and Avango, D. 2016. Industrial heritage sites in Spitsbergen (Svalbard), South Georgia and the Antarctic Peninsula: Sources of historical information. *Polar Science*, 10(3): 433–440.

Hamilakis, Y. 2013. Afterword: Eleven theses on the archaeology of the aenses. In Day, J. (ed.), *Making Senses of the Past: Toward a Sensory Archaeology*. Southern Illinois University Press, Carbondale, pp. 409–419.

Hamilakis, Y. 2014. *Archaeology and the Senses: Human Experience, Memory, and Affect*. Cambridge University Press, New York.

Harrowfield, D.L. 1978. Historical archaeology in Antarctica. *New Zealand Archaeological Association Newsletter*, 21(3): 95–102.

Harrowfield, D.L. 1984. The effects of wind on some historic Antarctic Huts. *Fram: The Journal of Polar Studies*, 1(2): 471–486.

Harrowfield, D.L. 1988. Historic sites in the ross dependency, Antarctica. *Polar Record*, 24(151): 277–284.

Harrowfield, D.L. 1991. Archaeology of Borchgrevink's stores hut, Cape Adare, Antarctica. *New Zealand Journal of Archaeology*, 13: 177–197.

Harrowfield, D.L. 1996. The role of the wind in the destruction of an historic hut at Cape Adare, Antarctica. *Polar Record*, 32(180): 3–18.

Harrowfield, D.L. 2005. Archaeology on ice: A review of historical archaeology in Antarctica. *New Zealand Journal of Archaeology*, 26: 5–28.

Hattersley-Smith, G. 1991. *The history of place names in the British Antarctic Territory*. British Antarctic Survey Scientific Reports No. 113 (pt1).

Hayman, S., Hughes, J. and Lazer, E. 1998. *Deterioration Monitoring and Tourism Management at Cape Denison (Mawson's Huts), Australian Antarctic Territory*. Report to the Australian Heritage Commission.

Headland, R.K. 1984. *The Island of South Georgia*. Cambridge University Press, Cambridge.

Headland, R.K. 2009. *A Chronology of Antarctic Exploration*. Quartritch, London.

Headland, R.K. 2018a. Published literature from Antarctic sealers: First-hand accounts. In Headland, R.K. (ed.), *Historical Antarctic Sealing Industry. Proceedings of an International Conference in Cambridge 16–21 September 2016*. Scott Polar Research Institute, Occasional Publication Series, Cambridge, pp. 11–33.

Headland, R.K. 2018b. Sealers wintering in the South Shetland Islands. In Headland, R.K. (ed.), *Historical Antarctic Sealing Industry. Proceedings of an International Conference in Cambridge 16–21 September 2016*. Scott Polar Research Institute, Occasional Publication Series, Cambridge: 147–149.

Hemmings, A.D., Roberts, P. and Dodds, K. (eds.). 2017. *Handbook on the Politics of Antarctica*. Edward Elgar, Cheltenham.

Henderson, C. 2009. The search for Mawson's air tractor. In *Mawson's Huts Foundation. Mawson's Huts conservation expedition 2008–2009 – Expedition Report*. Unpublished report for the Mawson's Huts Foundation, pp. 69–123.

Henderson, C. 2010. The search for Mawson's air tractor. *Australian Antarctic Magazine*, 18: 30–32.

Hince, B. 2000. *The Antarctic Dictionary: A Complete Guide to Antarctic English*. CSIRO Publishing and Museum Victoria, Collingwood.

Hissa, S. 2012. *Tempo e Arqueologia: Experiências Materiais e Imateriais de Antártica*. Master's diss. Universidade Federal de Minas Gerais, Belo Horizonte.

Hobbs, G.J. 1968. *The Geology of the South Shetland Islands IV: The Geology of Livingston Island*. British Antarctic Survey Scientific Reports, No 47.

Holmes, L. 1861. *The Arctic Whaleman: Or, Winter in the Arctic Ocean: Being a Narrative of the Wreck of the Whale Ship Citizen. Together With a Brief History of Whaling*. Thayer & Eldridge, Boston.

Howard, B. 1940. *Rakiura: A History of Stewart Island, New Zealand*. Reed, Dunedin.

Howes, D. 2006. Scent, sound and synesthesia. Intersensoriality and material culture theory. In Tilley, C., Keane, W., Küchler, S., Rowlands, M. and Syper, P. (eds.), *Handbook of Material Culture*. Sage, Trowbridge, pp. 161–172.

Howgego, R.J. 2003. *Encyclopedia of Exploration to 1800*. Hordern House, Sydney.

Howkins, A. 2016. *The Polar Regions. An Environmental History*. Polity Press, Cambridge.

Howkins, A. 2017. *Frozen Empires: An Environmental History of the Antarctic Peninsula*. Oxford University Press, Oxford.

Hughes, J. 1986. *Mawson's Huts Interim Conservation Report*. Report to Project Blizzard, Vol. 6.

Hughes, J. 2011. *Deterioration Processes Affecting Historic Sites in Antarctica and the Conservation Implications*. PhD diss. University of Canberra, Canberra.

Hughes, J. and Lazer, E. 2000. Importance of 'historic sites' on Heard Island for protection of scientific resources and environmental management of a World Heritage site'. In *Papers and Proceedings of the Royal Society of Tasmania*, 133(2): 71–77.

International Polar Heritage Committee. 2022. *ICOMOS Antarctic Archaeology Guidelines*. Retrieved from https://www.iphc-icomos.org/polar-heritage-resources/icomos-antarctic-archaeology-guidelines

Johnson, M. 1996. *An Archaeology of Capitalism*. Blackwell, Oxford.

Johnson, M. 1999. Rethinking historical archaeology. In Funari, P., Hall, M. and Jones, S. (eds.), *Historical Archaeology. Back from the Edge*. Routledge, New York, pp. 23–36.

Jones, A.G.E. 1985. British sealing in New South Shetland 1819–26. Part I. *Great Circle*, 7(1): 9–22; Part II. *Great Circle*, 7(2): 74–87.

Jones, K.L. 2009. Terror Cove, Port Ross: The German Transit of Venus Expedition 1874 and other scientific and discovery expeditions from 1840 to the start of World War I. In Dingwall, P.R, Jones, K.L. and Egerton, R. (eds.), *In Care of the Southern Ocean: An Archaeological and Historical Survey of the Auckland Islands*. New Zealand Archaeological Association Monograph, Auckland, pp. 87–103.

Kirker, J. 1970. *Adventures to China: Americans in the Southern Oceans 1792–1812*. Oxford University Press, New York.

Kittel, P. 2001. Inventory of whaling objects on the Admiralty Bay shores (King George Island, South Shetland Islands) in the years 1996–1998. *Polish Polar Research*, 22(1): 45–70.

Larson, E.J. 2011. *An Empire of Ice: Scott, Shackleton, and the Heroic Age of Antarctic Science*. Yale University Press, New Haven.

Latour, B. 1994. *Jamais Fomos Modernos: Ensaio de Antropologia Simétrica*. Editora 34, Rio de Janeiro.

Latour, B. 2012. *Reagregando o Social. Uma Introdução à Teoria do Ator-Rede*. EDUFBA e EDUSC, Salvador.

Lazer, E. 1985. *Recommendations for Future Archaeological and Conservation Work at the Site Associated with Mawson's Hut, Commonwealth Bay, Antarctica*. Unpublished report for Project Blizzard and the Australian Heritage Commission.

Lazer, E. 1998. *Reports on the AAP Mawson's Huts Foundation Expedition to Cape Denison, December 1997-February 1998*. Unpublished Report to Australian Antarctic Division and AAP Mawson's Huts Foundation.

Lazer, E. and McGowan, A. 1990. *Heard Island Archaeological Survey (1986–1987), revised 2nd edition*. University of Sydney, Sydney.

Leane, E. 2012. *Antarctica in Fiction: Imaginative Narratives of the Far South*. Cambridge University Press, Cambridge.

Leane, E. and McGee, J. 2020. Anthropocene Antarctica: Approaches, issues and debates. In Leane, E. and McGee, J. (ed.), *Anthropocene Antarctica. Perspectives from the Humanities, Law and Social Sciences*. Routledge, New York, pp. 1–14.

Ledingham, R.B. 1979. Expedition to renovate the 1912–13 Australian Antarctic Expedition Base Hut. *Polar Record*, 19(122): 485–492.

Le Mouël, J-F. 1994. *Rapport de la Mission du Patrimoine aux Iles Kerguelen (Campagne 1994)*. Unpublished report to Mission du Patrimoine, Terres Australes et Antarctiques Francaises.

Le Mouël, J-F. 1996. *Rapport de la Mission du Patrimoine aux Iles Kerguelen, Saint-Paul et Amsterdam (Campagnes 1995–1996)*. Unpublished report for Territoire des Terres Australes et Antarctiques Francaises.

Le Mouël, J-F. 2004. Heritage in the French Sub-Antarctic territory: Between urgency and emergency. In Barr, S. and Chaplin, P. (eds.), *Cultural Heritage in the Arctic and Antarctic Regions*. International Polar Heritage Committee of ICOMOS, Monuments and Sites VIII, Oslo, pp. 60–64.

Le Mouël, J-F. 2006. *Operation Obelix; Crozet – Ile de la Possession – Sauvetage d'un Fondoir de Phoquiers – Baleiniers*. Unpublished report for Territoire des Terres Australes et Antarctiques Francaises.

Le Mouël, J-F., Arnaud, P., Courbon, P., Rebiere, J., Moreignaux, N. and Jacquot, E. 2013. La vallée des Phoquiers aux îles Crozet. Un fondoir à graisse, témoin des premières occupations humaines. *Archéopages: Archéologie et Société*, 38: 40–47.

Leone, M. 1988. The Georgian Order as the order of merchant capitalism in Annapolis, Maryland. In Leone, M. and Potter, P. (ed.), *The Recovery of Meaning Historical Archaeology in the Eastern United States*. Smithsonian Institution Press, Washington, 235–261.

Lewis-Smith, R.I. 2018. An early 19th century sealers' structure at Yankee Harbour, Greenwich Island, South Shetland Islands. In Headland, R.K. (ed.), *Historical Antarctic Sealing Industry. Proceedings of an International Conference in Cambridge 16–21 September 2016*. Scott Polar Research Institute, Occasional Publication Series, Cambridge, pp. 151–158.

Lewis-Smith, R.I. and Simpson, H.W. 1987. Early nineteenth century sealers refuges on Livingston Island, South Shetland Islands. *British Antarctic Survey Bulletin*, 74: 48–72.

Llanos, N. 2019. Housewives at the end of the world. Chilean women living in Antarctica, 1984–1986. *Polar Journal*, 9(2): 358–370.

Lucero, V. and Stehberg, R. 1996. Arqueología histórica Antártica: Relevamiento y excavación de sitios de cazadores de lobos del siglo pasado. Islas Shetland del Sur. *Historical Archaeology in Latin America*, 14: 99–106.

Lüdecke, C. 2004. German meteorological and magnetic base station on the Kerguelen Islands (1901–1903). In Barr, S. and Chaplin, P. (eds.), *Cultural Heritage in the Arctic and Antarctic Regions*. International Polar Heritage Committee of ICOMOS, Monuments and Sites VIII, Oslo, pp. 65–68.

McGowan, A. 1987. Archaeology from the ice: Excavation methods in a frozen hut. *Australian Historical Archaeology*, 5: 49–53.

McGowan, A. 1988. Historical archaeology at Cape Denison, Commonwealth Bay, Antarctica. *Polar Record*, 24(149): 101–110.

McGowan, A. 2000. On their own: Towards an analysis of sealer's sites at Heard Island. *Papers and Proceedings of the Royal Society of Tasmania*, 133(2): 61–70.

McGowan, A. and Lazer, E. 2018. 19th century sealers' sites on Heard Island. In Headland, R.K. (ed.), *Historical Antarctic Sealing Industry. Proceedings of an International Conference in Cambridge 16–21 September 2016.* Scott Polar Research Institute, Occasional Publication Series, Cambridge, pp. 49–60.

McKenzie, J. 1992. Heroic myths of empire. In McKenzie, J. (ed.), *Popular Imperialism and the Military 1850–1950.* Manchester University Press, New York.

Maddison, B. 2014. *Class and Colonialism in Antarctic Exploration, 1750–1920.* Pickering and Chatto, London.

Management Plan for Antarctic Specially Protected Area No. 126 Byers Peninsula, Livingston Island, South Shetland Islands. ATCM XLIV (2022) Measure 10 (https://www.ats.aq/devAS/Meetings/Measure/760?s=1&from=1/1/1958&to=1/1/2158&cat=0&top=40&type=0&stat=0&txt=&curr=0&page=1, accessed September 26, 2022).

Martin, F., Morello Repetto, F., Alfonso-Durruty, M., Massone, M. and San Roman, M. 2015. Archaeology of Fueguian Islands: Tierra del Fuego, Dawson and Navarino, human settlement and cultural interaction (Patagonia, Chile). Work presented at the 80th Annual Meeting of the Society for American Archaeology, San Francisco, California. Retrieved from https://core.tdar.org/document/398028/archaeology-of-fueguian-islands-tierra-del-fuego-dawson-and-navarino-human-settlement-and-cultural-interaction-patagonia-chile

Martín-Bueno, M. 1995a. Arqueología Antártica: El Proyecto San Telmo y el descubrimiento de la Terra Australis Antártica. *Actas del V Simposio de Estudios Antárticos.* Comisión Interministerial de Ciencia y Tecnología, Madrid, pp. 421–428.

Martín-Bueno, M. 1995b. *Proyecto San Telmo. Informe sobre las Actividades Científicas de España en Antártida durante la Campaña 1993–94.* Comisión Interministerial de Ciencia y Tecnología, Madrid, pp. 249–265.

Martín-Bueno, M. 1996. *Proyecto de San Telmo: Arqueología Terrestre y Subacuática en Isla Livingston e Isala Desolación (Antártida). Informe sobre las Actividades Científicas de España en Antártida durante la Camppaña 1994–95.* Comisión Interministerial de Ciencia y Tecnología, Madrid, pp. 173–179.

Martín-Cancela, E. 2018. *Tras las Huellas del San Telmo: Contexto, Historia y Arqueología en la Antártida.* Prensas de la Universidad de Zaragoza, Zaragoza.

Mawer, G.A. 2006. *South by Northwest: The Magnetic Crusade and the Contest for Antarctica.* Wakefield Press, Kent Town.

Mawson's Huts Foundation. 2008. *Mawson's Huts Conservation Expedition 2007–2008: Expedition Reports.* Unpublished report for the Mawson's Huts Foundation.

Mawson's Huts Foundation. 2010. *Mawson's Huts Foundation 2009–2010 Expedition Reports.* Unpublished report for the Mawson's Huts Foundation.

Mawson's Huts Foundation. 2011. *Mawson's Huts Foundation 2010–2011 Expedition Reports.* Unpublished report for the Mawson's Huts Foundation.

Mayorga, M. 2016. Antecedentes históricos referidos a la caza de Lobos Marinos y su interacción con el medio geográfico y humano en el extremo Austral Americano: El caso del lobero escocés William Low. *Magallania*, 44(2): 37–64.

Mayorga, M. 2020. *Pieles, Tabaco y Quillangos: Relaciones entre Loberos Angloestadounidenses y Aborígenes Australes en la Patagonia (1780–1850).* Ediciones de la Subdirección de Investigación, Santiago.

Melville, H. 1851. *Moby-Dick, or The White Whale.* Harper & Brothers, London.

Merleau-Ponty, M. 1993. *Fenomenología de la Percepción*. Planeta Agostini, Buenos Aires.

Merleau-Ponty, M. 1977. *El Ojo y el Espíritu*. Paidós, Buenos Aires.

Moreno, P. 2000. Botellas de vidrio en la Península Byers, Isla Livingston, Shetland del Sur. *Actas del III Congreso Argentino de Americanistas*. Sociedad Argentina de Americanistas, Buenos Aires, pp. 207–228.

Mota, M. 2018. *Entre a Pena e o Papel: Uma Análise Material de Textos*. BA diss. Universidade Federal de Minas Gerais, Belo Horizonte.

Motos, L. and Wilson, D. 2006 *The Knowledge Base for Fisheries Management*. Elsevier, Amsterdam.

Motte, V. and Vidal, J. 2020. Archipel Crozet – Vallée des phoquiers. *ADLFI. Archéologie de la France. Informations, Océan Indien et TAAF*. Retrieved from https://journals.openedition.org/adlfi/37964

Moulton, W. 1804. *A Concise Extract from the Sea Journal of William Moulton: Written on Board the Onico, in a Voyage from the Port of New London in Connecticut to Staten Island in the South Seas*...The author, Utica.

Muñoz, A. 1997. El registro arqueofaunístico de la Península Byers, Isla Livingston, Shetland del Sur. *Actas de las IV Jornadas sobre Investigaciones Antárticas*, vol. II. Instituto Antártico, Buenos Aires, pp. 11–15.

Muñoz, S. 2000. Zooarqueología de la Isla Livingston, Shetland del Sur. *Archaeofauna*, 9: 39–57.

Myers, N. 2016. Longitudinal Antarctica: A continent in the abstract. *Cartographica*, 5(1): 38–50.

Nævestad, D., Basberg, B. and Rossnes, G. 1996. Industrial archaeology at South Georgia: Methods and results. *Polar Record*, 32(180): 51–66.

Nash, M. (ed.). 2009. *Captain Robinson: The Reminiscences of a Tasmanian Master Mariner James William Robinson 1824–1906*. Blubber Head Press, Hobart.

Nash, M. 2000. *Conserving Macquarie Island's Public Historic Heritage: Conservation Planning Guidance*. Unpublished report for the Cultural Heritage Branch, Department of Primary Industries, Water and Environment, Tasmania.

Neil, A. 2019. South Georgia's sealing past uncovered. *Newsletter of the South Georgia Heritage Trust and the Friends of South Georgia Island*, April: 1–5.

Nolasco, R. 2018a. Buscando simetrias no fim do mundo: pessoas, mamíferos marinhos e objetos na antártica do século XIX. *Revista Latino-Americana de Arqueologia Histórica*, 12(1): 101–121.

Nolasco, R. 2018b. *Pessoas, Mamíferos Marinhos e Objetos: Um Olhar Simétrico sobre a Antártica do Século XIX*. Master's diss. Universidade Federal de Minas Gerais, Belo Horizonte.

Nordhoff, C. 1856. *Whaling and Fishing*. Moore, Wilstach, Keys & Co, Cincinnati.

Norman, F.I, Gibson, J.A.E. and Burgess, J.S. 1998. Klarius Mikkelsen's 1935 landing in the Vestfold Hills, East Antarctica: Some fiction and some facts. *Polar Record*, 34(191): 293–304.

O'Gorman, F. 1963. The return of the Antarctic fur seal. *New Scientist*, 20: 374–6.

Olmsted, F. 1841. *Incidents of a Whaling Voyage, to which are Added Observations on the Scenery, Manners and Customs, and Missionary Stations of the Sandwich and Society Islands*. D. Appleton & Co., New York.

Orquera, L. and Piana, E. 1995. Túnel VII en la secuencia arqueológica del Canal Beagle: Hipótesis y expectativas de los investigadores argentinos. In Estévez, J. and Vila Mitjá, A. (eds.), *Encuentros en los Conchales Fueguinos. Traballs d'Etnoarqueologia 1*. CSIC, Barcelona, pp. 25–46.

Orser, C. 1996. *A Historical Archaeology of the Modern World*. Plenum Press, New York.

Parks and Wildlife Service. 2006. *Macquarie Island Nature Reserve and World Heritage Area Management Plan 2006*. Parks and Wildlife Service, Department of Tourism, Arts and the Environment, Tasmania.

Payró, R. 1898. *La Australia Argentina. Excursión Periodística á las Costas Patagónicas, Tierra del Fuego é Isla de los Estados*. Imprenta de la Nación, Buenos Aires.

Pearson, M. 1986. *Report for the Antarctic Historic Sites and Monuments Advisory Committee on a Vvisit to Mawson's Huts, Commonwealth Bay, 25–29 December 1986*. Report to the Antarctic Historic Sites and Monuments Advisory Committee and the Australian Heritage Commission.

Pearson, M. 1992. Expedition huts in Antarctica -1899–1917. *Polar Record*, 28(167): 261–276.

Pearson, M. 1993. *Mawson's Huts Historic Site Conservation Management Plan*. Unpublished report for the Australian Antarctic Division and the Mawson's Huts Conservation Committee.

Pearson, M. 2004. Artefact or rubbish—A dilemma for Antarctic managers. In Barr, S. and Chaplin, P. (eds.), *Cultural Heritage in the Arctic and Antarctic Regions*. International Polar Heritage Committee of ICOMOS, Monuments and Site VIII, Oslo, pp. 39–43.

Pearson, M. 2011. Polar heritage conservation and archaeology. In Barr, S. and Chaplin, P. (eds.), *Polar Settlements—Location, Techniques and Conservation*. International Polar Heritage Committee of ICOMOS International Council on Monuments and Sites, Oslo, pp. 24–34.

Pearson, M. 2018a. Living under their boats: A strategy for southern sealing in the nineteenth century — Its history and archaeological potential. *The Polar Journal*, 8(1): 68–83.

Pearson, M. 2018b. Australians sealers in the Antarctic Region -1820–22. *The Great Circle, Journal of the Australian Association for Maritime History*, 40(2): 105–121.

Pearson, M. 2018c. Sealing in the southern oceans: Towards an archaeological history. In Headland, R.K. (ed.), *Historical Antarctic Sealing Industry. Proceedings of an International Conference in Cambridge 16–21 September 2016*. Scott Polar Research Institute, Occasional Publication Series, Cambridge, pp. 35–47.

Pearson, M. 2018d. Nineteenth century sealing vessels in the southern oceans — An introduction to their history, wrecks and archaeological potential. *Australasian Journal of Maritime Archaeology*, 42: 25–38.

Pearson, M. 2020. 'Knowing' the South Shetlands: The role of sealer's charts. *Shima*, 14(1): 108–132.

Pearson, M. and Lennon, J. 2010. *Pastoral Australia: Fortunes, Failures and Hard Yakka*. CSIRO Press, Melbourne.

Pearson, M. and Salerno, M. (in press). Stateless heritage – the sealing sites of the South Shetland Islands, Antarctica. In Stiefel, B.L. and Peleg, S-A. (eds.), *Heritage without boundaries: Yearbook of Transnational History 6*. Fairleigh Dickenson University/Rowman & Littlefield.

Pearson, M. and Senatore, M.X. 2014. Historical archaeology in Antarctica. In Smith, C. (ed.), *Encyclopedia of Global Archaeology*. Vol I. Springer, New York, pp. 269–272.

Pearson, M. and Stehberg, R. 2006. Nineteenth century sealing sites on Rugged Island, South Shetland Island. *Polar Record*, 42(223): 335–347.

Pearson, M. and Stehberg, R. 2011. Geographic and technological influences on the location, nature and conservation of nineteenth century sealing sites in the South Shetland Islands. In Barr, S. and Chaplin, P. (eds.), *Polar Settlements—Location, Techniques and Conservation*. International Polar Heritage Committee of ICOMOS, International Council on Monuments and Sites, Fjellhamar, pp. 86–93.

Pearson, M., Stehberg, R., Zarankin, A., Senatore, M.X. and Gatica, C. 2008. Sealer's sledge excavated on Livingston Island, South Shetland Islands. *Polar Record*, 44(4): 362–364.

Pearson, M., Stehberg, R., Zarankin, A., Senatore, M.X. and Gatica, C. 2010. Conserving the oldest historic sites in the Antarctic: The challenges in managing the sealing sites in the South Shetland Islands. *Polar Record*, 46(1): 57–64.

Pearson, M., Zarankin, A. and Salerno, M. 2020. Exploring and exploiting Antarctica; the early human interactions. In Oliva, M and Fernandez, J. (eds.), *Past Antarctica: Paleoclimatology and Climate Change*. Academic Press, London, pp. 259–278.

Peat, N. 2013. *Shackleton's Whisky: The Extraordinary Story of an Heroic Explorer and Twenty-Five Case of Unique MacKinlay's Old Scotch*. Cornerstone, London.

Pena, W.L.S. 2013. *Problemas de Gênero em Alto Mar: A Construção de Masculinidade(s) em Baleeiros da Nova Inglaterra, Século XIX*. BA diss. Universidade Federal de Minas Gerais, Belo Horizonte. Ms.

Pena, W.L.S. 2016. *Negociando Gêneros com o Paralelo 60: Por uma Genealogia do Prestígio Antártico*. Mg. diss. Universidade Federal de Minas Gerais, Belo Horizonte. Ms.

Petchey, P., Egerton, R. and Boyd, W. 2015. A Spanish man-o-war in New Zealand? The 1864 wreck of Grafton and its lessons for pre-Cook shipwreck claims. *International Journal of Nautical Archaeology*, 44: 362–370.

Petersen, R. 2021. Lessons in temporal permanence. McMurdo Station. In Foscari, G. (ed.), *Antarctic Resolution*. Lars Müller Publishers, Zurich, pp. 453–455.

Phelps, W.D. 1871. *Fore and Aft; or, Leaves from the Life of an Old Sailor, by Webfoot*. Nichols & Hall, Boston.

Pinochet de la Barra, O. 1991. El misterio del San Telmo. ¿Náufragos españoles pisaron por primera vez la Antártida? *Revista Historia*, 16(186): 31–37.

Plan de Manejo de la Reserva Provincial Isla de Los Estados, Islas de Año Nuevo e Islotes Adyacentes. n.d. Retieved from https://desarrollosustentable.tierradelfuego.gob.ar

Prickett, N. 2009. Sealing in the Auckland Islands. In Dingwall, P.R., Jones, K.L. and Egerton, R. (eds.), *In care of the Southern Ocean: An archaeological and historical survey of the Auckland Islands*. New Zealand Archaeological Association Monograph 27: 39–57.

Quartermain, L.B. 1961. Restoration of the Scott and Shackleton Huts on Ross Island 1960–61. *Polar Record*, 10: 605–608.

Quartermain, LB. 1963. *Two Huts in the Antarctic*, NZ Government Printer, Wellington.

Rando, S. and Davies, M. 1996. *Davis Station Heritage Study, Parts A & B*. Unpublished report for the Australian Antarctic Division.

Reinhard, A. 2018. *Archaeogaming: An Introduction to Archaeology in and of Video Games*. Berghahn Books, Oxford.

Richards, R. 1991. The commercial exploitation of sea mammals at Iles Crozet and Prince Edward Island before 1850. *Polar Monograph*, 1: 1–19. (Re-published as 'Sealing at Prince Edward Island, Marion and Iles Crozet' in Richards 2010: 23–61.)

Richards, R. 2003. New Market Evidence on the depletion of southern fur seals: 1788–1833. *New Zealand Journal of Zoology*, 30(1): 1–9.

Richards, R. 2010. *Sealing in the Southern Oceans 1788–1833*. Paremata Press, Wellington.

Richards, R. and Knight, L. (eds.). 1996. *Jorgen Jorgenson's observations on Pacific Trade; and Sealing and Whaling in Australia and New Zealand Waters before 1805*. The Paremata Press, Wellington.

Richardson, L-J. 2009. *Measuring the Success of Digital Public Archaeology: Towards a Model for Public Participation in Internet Archaeology*. Master's diss. University of London, London.

Riffenburgh, B. 1994. *The Myth of the Explorer: The Press, Sensationalism, and Geographical Discovery*. Oxford University Press, Oxford.

Riffenburgh, B. (ed). 2007. *Encyclopedia of the Antarctic*. Routledge, London.

Ritchie, N.A. 1988. *Ross Island Historic Huts, Antarctica, Report on archaeological heritage work and future management considerations for the Antarctic Heritage Trust*. Department of Conservation, New Zealand Science and Research Internal Report No. 15.

Ritchie, N.A. 1989. *Ross Island Historic Huts, Antarctica, Report on archaeological and restoration work and future management considerations, Event K281 1988–89, for the Antarctic Heritage Trust*. Department of Conservation, New Zealand, Science and Research Internal Report No. 47.

Ritchie, N.A. 1990a. Archaeological techniques and technology on Ross Island, Antarctica. *Polar Record*, 26(159): 257–264.

Ritchie, N.A. 1990b. 78 South, 166 East: Archaeology in the Ross Sea Dependency, Antarctica. *Australian Historical Archaeology*, 8: 23–27.

Ritchie, N.A. 2005. *Archaeological Work Cape Evans & Cape Royds, Ross Island, Antarctica: January 2005, part of Event K440*. Unpublished report for the Antarctic Heritage Trust.

Ritchie, N.A. 2006. Frozen solid: Recent archaeological work at Shackleton's Cape Royds hut site (1907–09), Ross Island, Ross Dependency, Antarctica. *Australasian Historical Archaeology*, 24: 31–38.

Ritchie, N.A. 2008. *Archaeological Work Cape Evans Ross Island, Antarctica: November-December 2007, Part of Event K170*. Unpublished report for the Antarctic Heritage Trust.

Ritchie, N.A. and Fyfe, R. 1995. *Report on Archaeological and Conservation Related Fieldwork (Event K282) in the Ross Dependency during the 1994–95 Field Season*. Report for the Antarctic Heritage Trust.

Ritchie, N.A. and Simmons, A.L. 1987. *Ross Island Historic Huts Preservation: A Report on Archaeological Work and Management Considerations: Event K072B, 1986–87*. Unpublished report, New Zealand Historic Places Trust, Cromwell.

Robb, A. 1988. Archaeological site survey on Heard Island: North West Coast. In Antarctic Division, *1987–88 Australian Antarctic Research Program: Initial Summary of Research Activity*. Antarctic Division, Department of Science, Hobart: 155–156.

Roberts, B.B. 1958a. Chronological list of Antarctic expeditions. *Polar Record*, 9(59): 97–134.

Roberts, B.B. 1958b. Chronological list of Antarctic expeditions. *Polar Record*, 9(60): 191–239.

Ronne, E. 2004. *Antarctica's First Lady*. Clifton Steamboat Museum and Three Rivers Council, Beaumont.

Roura, R.M. 2017. Antarctic cultural heritage: Geopolitics and management. In Dodds, K., Hemmings, A. and Roberts, P. (eds.), Handbook of Antarctic Politics. Edward Elgar Press, Cheltenham, pp. 468–484.

Rosa, L. *Lembranças de um Mundo Vazio: Arqueologia da Descoberta e Ocupação Inicial do Continente Antártico*. Master's diss. Universidade Federal de Minas Gerais, Belo Horizonte.

Salerno, M. 2006. *Arqueología de la Indumentaria: Prácticas e Identidad en los Confines del Mundo Moderno*. Del Tridente, Buenos Aires.

Salerno, M. 2007. 'Identidades extremas': moda, vestido e identidad en los confines de la sociedad moderna (Antártida, siglo XIX). *Arqueología*, 13: 185–211.

Salerno, M. 2009. Zapatos rotos: Una aproximación al calzado en arqueología histórica. *Actas del VI Congreso Argentino de Americanistas*. Vol. II. Dunken, Buenos Aires, pp. 369–383.

Salerno, M. 2011. *Persona y Cuerpo-Vestido en la Modernidad: Un Enfoque Arqueológico*. PhD diss. Universidad de Buenos Aires, Buenos Aires. Ms.

Salerno, M. 2014. Sealers were not born but made: Sensory-motor habits, subjectivities and 19th-century voyages to the South Shetland Islands. In Pellini, J.R., Zarankin, A. and Salerno, M.A. (eds.), *Coming to Senses: Topics in Sensory Archaeology*. Cambridge University Press, Cambridge, pp. 77–104.

Salerno, M. 2015. Persona y cuerpo-vestido en la modernidad. Los loberos-balleneros de la industria capitalista del siglo XIX. *Vestigios. Revista Latino-Americana de Arqueologia Histórica*, 9(1): 113–153.

Salerno, M. and Cruz, J. 2019. Between words and oceans. Logbooks and the Antarctic sealing industry. Work presented at SC-HAAS Conference 2019 – Antarctic Connections at the End of the World: Understanding the Past and Shaping the Future. Scientific Committee on Antarctic Research, Ushuaia, 3–5 April.

Salerno, M., Cruz, J., Rigone, R. and Zarankin, A. 2021. Making visible the invisibles. Discussing new approaches to the diversity among the nineteenth century sealing crews. Work presented at 2021SCAR SC-HASS Biennal Conference The Global Antarctic. Scientific Committee on Antarctic Research, Kobe, 18–19 November.

Salerno, M., Cruz, J. and Zarankin, A. 2019. Inside or outside capitalism? Sealers' lives, food, and clothing onboard sealing vessels and on Antarctic hunting grounds. In Nyman, J., Fogle, K. and Beaudry, M. (eds.), *Historical Archaeology of Shadow and Intimate Economies*. University Press of Florida, Gainesville, pp. 158–177.

Salerno, M., Cruz, J. and Zarankin, A. 2021. Sealing at the end of the world. Exploring the connections between fur seal exploitation in Antarctica and Tierra del Fuego. Work presented at 2021 *SCAR SC-HASS Biennial Conference*. Scientific Committee on Antarctic Research, Kobe, 18–19 November.

Salerno, M., Rigone, R. and Zarankin, A. 2018. Explorando bitácoras. Aproximaciones al accionar de loberos y balleneros en Tierra del Fuego durante el siglo XIX. Work presented at *VII Congreso Nacional de Arqueología Histórica*. Universidad Nacional de Rosario, Rosario, 22–26 October.

Salerno, M. y Rodríguez, S.M. 2007. 'Historias preservadas': Arqueología histórica, conservación y una propuesta integradora. El caso de la colección textil antártica. *Fuentes e Interdisciplina. Actas de las II Jornadas Multidisciplinarias*. Dunken, Buenos Aires, pp. 349–358.

Salerno, M., Rodriguez, S.M. and Valladares, I.V. 2007. Memorias materiales: Conservación y protección de un pasado 'sin historia'. In Vazquez, C. and Palacios, O. (eds.), *Patrimonio Cultural: La Gestión, el Arte, la Arqueología y las Ciencias Exactas Aplicadas*. Eudeba, Buenos Aires, pp. 235–244.

Salerno, M.A. and Zarankin, A. 2014. En busca de las experiencias perdidas. Arqueología del encuentro entre los loberos y las Islas Shetland del Sur (Antártida, Siglo XIX). *Vestigios. Revista Latino-Americana de Arqueologia Histórica*, 8(1): 131–157.

Salerno, M.A., Zarankin, A., Cruz, M.J., Rigone, R.C. and Weissel, M. 2021. Explotación lobera en Tierra del Fuego: Primeras aproximaciones desde la arqueología histórica. *Teoría y Práctica de la Arqueología Histórica Latinoamericana*, 13(1): 61–80.

Salerno, M., Zarankin, A. and Senatore, M.X. 2010. La visión cartográfica: Expansión territorial y poder en el mundo moderno; el caso de las Isla Shetland del Sur (Antártida, Principios del Siglo XIX). In Patiño, D., Zarankin, A. and Mantilla, C. (eds.), *Arqueologías Históricas, Patrimonios diversos*. Universidad del Cauca, Cauca, pp. 15–32.

Schiavini, A. 1990. *Estudio de la Relación entre el Hombre y los Pinnípedos en el Proceso Adaptativo Humano al Canal Beagle, Tierra del Fuego, Argentina*. PhD diss. Universidad de Buenos Aires, Buenos Aires.

Schiffer, M. 1987. *Formation Processes of the Archaeological Record*. University of New Mexico Press, Albuquerque.

Senatore, M.X. 2018. Turismo y sitios arqueológicos en las Islas Shetland del Sur, Antártida. *Pasos, Revista de Turismo y Patrimonio Cultural*, 16(1): 99–116.

Senatore, M.X. 2019. Archaeologies in Antarctica from nostalgia to capitalism: A Review. *International Journal of Historical Archaeology*, 23(3): 755–771.

Senatore, M.X. and Zarankin, A. 1996. Perspectivas metodológicas en arqueología histórica. *Páginas sobre Hispanoamérica Colonial. Sociedad y Cultura*, 3: 113–122.

Senatore, M.X. and Zarankin, A. 1997. Arqueología histórica en Antártida. Avances en la investigación. *Actas del II Congreso Argentino de Americanistas* Vol. II. Sociedad Argentina de Americanistas, Liga Naval, Buenos Aires, pp. 585–603.

Senatore, M.X. and Zarankin, A. 1999. Arqueología histórica y expansión capitalista. Prácticas cotidianas y grupos operarios en Peninsula Byers, Isla Livingston de las Shatleand del Sur. In Zarankin, A. and Acuto, F. (eds.), *Sed Non Sataita. Teoria Social en la Arqueología Latinoamericana Contemporánea*. Ediciones del Tridente, Buenos Aires, pp. 171–88.

Senatore, M.X. and Zarankin, A. 2012. Tourism and the invisible historic sites in Antarctica. In *ICOMOS 17th General Assembly Symposium*. ICOMOS, Paris, pp. 599–608. Retrieved from http://openarchive.icomos.org/id/eprint/1147/

Senatore, M.X. and Zarankin, A. 2014. Against the domain of master narratives: Archaeology and history in Antarctica. In Gnecco, C. and Langebaek, C. (eds.), *Against the Typological Tyranny in Archaeology: A View from South America*. Springer, New York, pp. 121–132.

Shanks, M. 2007. Symmetrical archaeology. *World Archaeology*, 39(4): 589–596.

Shears, J. and Hall, J. 1993. Abandoned stations and field huts: The British approach to management. In Melander, O. and Fontana, L. (eds.). *Proceedings of the Fifth Symposium on Antarctic Logistics and Operations, San Carlos de Bariloche, Argentina, June 1992*. Dirección Nacional del Antártico, Buenos Aires, pp. 12–26. Retrieved from https://www.bas.ac.uk/data/our-data/publication/abandoned-stations-and-field-huts-the-british-approach-to-management/

Simpson, H.W. 1959. *'The discovery of fur sealers' relics in the South Shetland Islands.* British Antarctic Survey Archives Report, No FIDScB 195/59.

Siqueira, A. 2014. *Histórias Embriagadas: Uma Análise Diferenciada sobre Garrafas de Sítios Foqueiros na Antártica nos Séculos XVIII e XIX*. BA diss. Universidade Federal de Minas Gerais, Belo Horizonte.

Skinner, L-A. 2014. Archaeological excavations and artefact conservation at the heroic-era expedition bases, Ross Island, Antarctica. *Journal of Glacial Archaeology*, 1(1): 51–77.

Smith, I.W.G. 2002. *The New Zealand Sealing Industry: History, Archaeology and Heritage Management*. Department of Conservation, Wellington.

Smith, T.W. 1844. *A Narrative of the Life, Travels and Sufferings of Thomas W. Smith*, Wm.C. Hill, Boston.

Snape, I., Cole, C., Gore, D., Riddle, M. and Yarnell, M. 1998. *A Preliminary Assessment of Contaminants at the Abandoned Wilkes Station, East Antarctica, with Recommendations for Establishing an Environmental Management Strategy – October 1998*. Unpublished report for the Australian Antarctic Division.

Soares, F.C. 2020. *Mundos Virtuais: Uma Proposta de Teorização sobre a Prática das Simulações Computacionais na Arqueologia Antártica*. Posdoctoral Final report. Universidade Federal de Minas Gerais, Belo Horizonte.

Soares, F.C., Amorim, C.L.A. and Pena, W.L.S. 2019. Um fio de fumaca nos Mares do Sul: Cachimbos de caulim e masculinidades nas Islas Shetland do Sul (séculos XVIII e XIX). *Revista de Arqueologia*, 32(1): 129–159.

Soares, F.C. and Gardiman, G.G. 2017. Mais uma dose: Análise arqueobotânica do consumo de cerveja nas Shetland do Sul (Antártica). *Revista Habitus*, 15(2): 273–299.

Soares, F.C. and Martire, A. 2021. Transcendendo gretas da Antártica: Arqueologia Pública Digital e Archaeogaming nas Shetland do Sul. In Zarankin, A., Esteves Arantes, R.M., Rosa, L.H. and Soares, F.C (eds.), *Antártica em Minas Gerais: Avanços científicos nas áreas de Medicina/Fisiologia, Microbiologia e Arqueologia do Polo Sul e sua importância para o Brasil*. Imprensa Universitária da UFMG, Belo Horizonte, pp. 63–88.

Soares, F.C. and Mota, M. 2017. Arqueologia digital abaixo de zero: Uma proposta de mediação para a arqueologia antártica. *Vestígios. Revista Latino-Americana de Arqueologia Histórica*, 11(1): 19–39.

Soares, F.C., Nolasco, R.C. and Mota, M. 2018. Antarctic digital public archaeology. In Headland, R.K. (ed.), *Historical Antarctic Sealing Industry. Proceedings of an International Conference in Cambridge 16–21 September 2016*. Scott Polar Research Institute, Occasional Publication Series, Cambridge, pp. 139–146.

Soares, F.C., Rosa, L.A.S da, Joia, T.C. and Pena, W.L.S. 2016. A (des) Construção da Embriaguez em Solos Antárticos. In Soares, F.C. (ed.), *Comida, Cultura e Sociedade – Arqueologias da Alimentação no Mundo Moderno*. Editora Universirtaria UFPE, Recife.

Soares, F.C. and Zarankin, A. 2020. Sensing antarctica: Sensorial exhibit on antarctica at espaço do conhecimento UFMG, Brazil. *Revista Arqueologia Pública*, 14(1): 182–198.

South Georgia Association. 2021. YouTube South Georgia Association talk 20 May, 2021. Archaeology Fieldwork on South Georgia. Retrieved from https://www.youtube.com/watch?v=j3IKqjNT658&list=PL9dkbAGGDnDIhsih5rPXjOTS1r-rkDG0E&index=6

Spude, C. and Spude, R. 1993. *East Base. Historic Monument. Stonington Island, Antarctic Peninsula*. United Sates Department of the Interior, Denver.

Stackpole, E.A. 1953. *The Sea-Hunters: The New England Whalemen during Two Centuries 1635–1835*. Bonanza Books, New York.

Stackpole, E.A. 1955. *The Voyage of the Huron and the Huntress: The American Sealers and the Discovery of the Continent of Antarctica*. Marine Historical Association, Mystic.

Staples, M. 1996. *Conservation Sssessment of Selected Old Buildings and Artefacts, Mawson Station, Antarctica*. Unpublished report for the Australian Antarctic Division.

Starke, J. (ed.). 1986. *Journal of a Rambler: The Journal of John Boultbee*. Oxford University Press, Melbourne.

Stehberg, R.L. 1983a. Terra Australia Incógnita: Una ruta de investigación arqueológica. *Serie Científica del Instituto Antártico Chileno*, 30: 77–86.

Stehberg, R.L. 1983b. En torno a la autenticidad de las puntas de proyectil aborígenes descubiertas en las Isla Shetland del Sur. *Boletín Antártico Chileno*, 3: 21–22.

Stehberg, R.L. 1997. Pruebas arqueológicas de la presencia de aborígenes del extremo sur americano en las actividades de caza de lobos finos desarrolladas en las islas Shetland del Sur a principios del Siglo XIX. *III Reunión de Hist. Antártica Iberoamericana*. Punta Arenas 1996. Instituto Antártico Chileno, Santiago, pp. 107–117.

Stehberg, R.L. 2002. Catastro del patrimonio cultural de la costa norte de la isla Rey Jorge, Antártica. In Barnes, J.B. (ed.), *VI Encuentro de Historiadores Antárticos Iberoamericanos*. Instituto Antártico Chileno, Santiago, pp. 31–50.

Stehberg, R.L. 2003. *Arqueología Histórica Antártica: Aborígenes Sudamericanos en los Mares Subantárticos en el Siglo XIX*. Centro de Investigaciones Diego Barros Arana, Santiago.

Stehberg, R.L. 2004. Archaeologists document historical heritage in the South Shetland Islands. In Barr, S. and Chaplin, P. (eds.), *Cultural Heritage in the Arctic and Antarctic Regions*. ICOMOS Monuments and Sites VIII, International Polar heritage Committee, Oslo, pp. 69–72.

Stehberg, R.L. and Blanco, J.F. 2002. ¿Artefactos indígenas o chisperos para fusil? Un problema arqueológico en Isla Desolación, Antártica Chilena. *Anales Instituto Patagonia*, 30: 175–181.

Stehberg, R.L. and Cabeza, A. 1984. Primera excavación de arqueología histórica Antártica. *Boletín Antártico Chileno*, 4: 15–17.

Stehberg, R.L. and Cabeza, A. 1987. Comienzos de la arqueología histórica antártica en el sitio Cuatro Pircas. *Revista Chilena de Antropología*, 6: 83–111.

Stehberg, R.L. and Gatica, C. 2001. Reconocimiento arqueólogico de la costa norte de Isla Rey Jorge. *Boletín Antártico Chileno*, 20(1): 7–9.

Stehberg, R.L. and Lucero, V. 1985a. Contexto arqueológico del hallazgo de restos humanos en Cabo Shirref, Island Livingston. *Serie Científica del Instituto Antártico Chileno*, 45: 59–66.

Stehberg, R.L. and Lucero, V. 1985b. Arqueología histórica de la Isla Desolación. Evidencias de coexistencia entre cazadores de lobo de origen europeo y aborígenes del extremo sur americano, en la segunda década del siglo pasado. *Serie Científica del Instituto Antártico Chileno*, 45: 67–88.

Stehberg, R.L. and Lucero, V. 1995. Evidencias de coexistencia entre cazadores de lobos y aborígenes fueguinos en isla Desolación, Shetland del Sur, Antártica, a principios del siglo XIX. *Serie Científica del Instituto Antártico Chileno*, 45: 67–88.

Stehberg, R.L. and Lucero, V. 1996. Excavaciones arqueológicas en Playa Yámana, Ccabo Shirreff, Isla Livingston, Shetland del Sur, Antártica. *Serie Científica del Instituto Antártico Chileno*, 46: 59–81.

Stehberg, R.L. and Nilo, L. 1983. Procedencia antártica inexacta de dos puntas de proyectil. *Boletín Antártico Chileno*, 3(1): 2–21.

Stehberg, R.L., Pearson, M. and Gatica, C. 2005. Hallazgos histórico-arqueológicos en Isla Rugged, Antartica. *Boletín Antártico Chileno*, 24(1): 7–11.

Stehberg, R., Pearson, M., Blanchette, R. and Jurgens, J.A. 2009. A further note on a sealer's sledge, discovered on Livingston Island, South Shetland Islands. *Polar Record*, 45(3): 275.

Stehberg, R., Pearson, M., Zarankin, A., Senatore, M.X. and Gatica, C. 2008. Protection and preservation of the oldest sites of the Antarctic: The case of Fildes Peninsula and Byers Peninsula in the South Shetland Islands. In Barr, S. and Chaplin, P. (eds.), *Historical Polar Bases: Preservation and Management*. ICOMOS, Monuments and Sites XIV, International Council on Monuments and Sites, Paris, pp. 85–93.

Stevenson, M.G., Madsen, A. and Maloney, E. (eds.). 1997. *The Anthropology of Community-Based Whaling in Greenland: A Collection of Papers Submitted to the International Whaling Commission*. University of Alberta Press, Alberta.

Summerson, R.M.V. 2013. *The Protection of Wilderness and Aesthetic Values in Antarctica*. PhD diss. The University of Melbourne, Melbourne.

Swedish Polar Research Secretariat. 2019. Posting on Snow Hill Island. Retrieved from https://www.polar.se/en/news/2019/antarctic-expedition-documents-cultural-heritage/

Taylor, N.W. 1929. *Life on a Whaler or Antarctic Adventures in the Isle of Desolation*. New London County Historical Society, Occasional Publications, Vol II, New London.

Tin, T., Hemmings, A. and Roura, R. 2008. Pressures on the wilderness values of the Antarctic Continent. *International Journal of Wilderness*, 14: 7–12.

Tønnessen, J.N. and Johnsen, A.O. 1982. *The Hhistory of Modern Whaling*. C. Hurst & Company, and Australian National University Press, Canberra.

Torres, D. 1992. ¿Cráneo indígena en cabo Shirreff?: Un estudio en desarrollo. *Boletín Antártico Chileno*, 11: 2–6.

Torres, D. 1999. Observations on ca.175-year old human remains from Antarctica (Cape Shirreff, Livingston Island, South Shetlands). *International Journal of Circumpolar Health*, 58: 72–83.

Townrow, K. 1989. *Survey and Excavation of Historic Sites on Macquarie Island. Occasional Paper No 20.* Department of Lands, Parks and Wildlife, Hobart.

Townrow, K. 1990. *Post-Excavation and Analysis of Archaeological Projects on Macquarie Island.* Unpublished report for the Antarctic Scientific Advisory Committee, Department of Geography and Environmental Studies, University of Tasmania.

Townrow, K. and Nash, M. 1990. Close shave from the Sub-Antarctic: An analysis of timber samples from archaeological sites on Macquarie Island. *Bulletin of the Australian Institute for Maritime Archaeology*, 14(2): 9–16.

Townrow, K. and Shaughnessy, P.D. 1991. Fur seal skull from sealer's quarters at Sandy Bay, Macquarie Island, Southern Ocean. *Polar Record*, 27(162): 245–248.

Turner, G.A. 1979. *A Strategy for the Preservation and Management of Historic Sites in the Ross Dependency, Antarctica.* Department of Lands and Survey, Wellington.

Turner, G.A. and Harrowfield, D.L. 1984. *Corporate Strategic Plan for Ross Island Historic Sites, Ross Dependency, Antarctica.* Report to Historic Sites Management Committee, Ross Dependency Research Committee.

Valadares, L. 2020. *Pisando Firme. Um estudo Geoarqueológico na Antártica.* BA diss. Universidade Federal de Mina Gerais, Belo Horizonte.

Van Niekerk, T. 2016. *An Archaeological Study of the Solglimt Shipwreck Survivor Camp on Sub-Antarctic Marion Island.* Master's diss. University of South Africa, Pretoria.

Vázquez, M. and Santiago, F. 2014. Explotación industrial de pinnípedos en la factoría de Bahía Thetis (Tierra del Fuego, Argentina): Una aproximación zooarqueológica. *Revista Chilena de Antropología*, 29, first semester: 130–135.

Vázquez, M., Santiago, F. and Zangrando, A. 2019. La factoría de Bahía Thetis, antecedentes históricos y tafonomía de camadas de huesos de pinnípedos. *Magallania*, 47(2): 151–173.

Verrill, A.H. 1916. *The Real Story of the Whaler. Whaling, Past and Present.* D. Appleton & Company, New York.

Villagran, X.S. and Schaefer, C.E.R. 2011. Geoarqueologia das primeiras ocupaçoes humanas na Antártica. *Vestígios. Revista Latino-Americana de Arqueologia Histórica*, 5(1): 115–136.

Villagran, X.S., Schaefer, C.E.R. and Ligouis, B. 2013. Living in the cold: Geoarchaeology of sealing sites from Byers Peninsula (Livingston Island, Antarctica). *Quaternary International*, 30: 1–16.

Vincent, R. 2004. *Macquarie Island Station, Buckles Bay: Macquarie Island Cultural Heritage Management Plan.* Prepared for the Australian Antarctic Division.

Watson, N. and Ussher, J. 2018. *Hillary's Antarctica: Adventure, Exploration and Establishing Scott Base.* Allen and Unwin, Auckland.

Webb, E. 1925. *Field Survey and Reduction of Magnetograph Curves.* Scientific reports, Australasian Antarctic Expedition, 1911–1914. Series B, v. 1, Terrestrial Magnetism, pt. 1. Government Printer, Sydney, pp. 54–56.

Weddell, J. 1827. *A Voyage towards the South Pole, Performed in the Years 1822–24.* Longman, Rees, Orme, Brown & Green, London.

Wehi, P.M., Scott, N.J., Beckwith, J., Rodgers, R.P., Gillies, T., van Uitregt, V. and Watene, K. 2021. A short scan of Māori journeys to Antarctica. *Journal or the Royal Society of New Zealand.* Retrieved from https://www.tandfonline.com/doi/full/10.1080/03036758.2021.1917633

West, C. 2005. *New Zealand Subantarctic Islands Research Strategy*, Department of Conservation (Te Papa Atawhai), Invercargill. Retrieved from http://gdsindexnz.org/wp-content/uploads/2015/10/New-Zealand-Subantartic-Islands-Research-Strategy.pdf

White, M. 1967. Manuscript field notes of 1966/67 season in the South Shetlands, transcribed by Robert Burton, 2019. British Antarctic Survey. Archives Reference number AD6/2H/1966/N6.

Witmore, C. 2007. Symmetrical archaeology: Excerpts of a manifesto. *World Archaeology*, 39(4): 546–562.

Yates, F. 2007. *A Arte da Memória*. Editora Unicamp, Campinas.

Zangrando, A.F. 2010. Coastal archaeology and hunter-gatherers in Southeastern Tierra del Fuego. *Journal of Island and Coastal Archaeology*, 5(2): 288–291.

Zarankin, A., Hissa, S., Salerno, M., Froner, Y-A., Alkmim, G., Assis, L. and Batista, A. 2011. Paisagens em branco: Arqueologia e antropologia antárticas - Avanços e desafios. *Vestígios. Revista Latino-Americana de Arqueología Histórica*, 5(2): 9–52.

Zarankin, A. and Salerno, M. 2014. The "Wild" Continent? Some Discussions on the Anthropocene in Antarctica. *Journal of Contemporary Archaeology*, 1(1): 114–118.

Zarankin, A. and Salerno, M. 2017. Antarctic archaeology: Discovering the history of the southernmost end of the world. In Symonds, J. and Herva, V-S. (eds.), *The Oxford Handbook of Historical Archaeology*. Online publication OUP.

Zarankin, A. and Salerno, M. 2018. So far, so close. Approaching experience in the study of encounter between sealers and the South Shetland Islands. In Roberts, P., van der Watt, L-M. and Howkins, A. (eds.), *Antarctica and the Humanities*. Palgrave McMillan UK, London, pp. 79–103.

Zarankin, A., Salerno, M. and Howkins, A. 2018. From the Antarctic to New England: Remembrance of sealing and sealers. In Headland, R.K. (ed.), *Historical Antarctic Sealing Industry. Proceedings of an International Conference in Cambridge 16–21 September 2016*. Scott Polar Research Institute, Occasional Publication Series, Cambridge, pp. 107–119.

Zarankin, A. and Senatore, M.X. 1999. 'Estrategias y tácticas' en el proceso de ocupación de la Antártida –Siglo XIX. In Espinosa, S. (ed.), *Desde el País de los Gigantes. Perspectivas Arqueológicas en Patagonia*, Vol. 1. Universidad Nacional de la Patagonia Austral, Río Gallegos, pp. 315–327.

Zarankin, A. and Senatore, M.X. 2000. Hasta el fin del mundo. Arqueología en las Islas Shetland del Sur. El caso de Península Byers, Isla Livingston. *Praehistoria*, 3: 219–236.

Zarankin, A. and Senatore, M.X. 2005. Archaeology in Antarctica: Nineteenth-century capitalism expansion strategies. *International Journal of Historical Archaeology*, 9(1): 43–56.

Zarankin, A. and Senatore, M.X. 2007. *Historias de un Pasado en Blanco: Arqueología Histórica Antártica*. Argumentum, Belo Horizonte.

Zarankin, A. and Soares, F.C. 2021. Invisible heritage: New technologies and the history of Antarctica's sealers group. In Cunha, F. and Rabassa, J. (eds.), *Festivals and Heritage in Latin America. Interdisciplinary dialogues on culture, identity and tourism*. Springer, Berlin, pp. 257–269.

Zarankin, A., Soares, F.C. and Martire, A. 2021. Arqueologia pública e novas tecnologias: O caso do projeto Paisagens em branco. *Revista Teoría y Práctica de la Arqueología Histórica Latinoamericana*, 12: 9–25.

APPENDIX

Sealing sites identified in the South Shetland Islands (to 2018)

Name	*Latitude/ longitude*	*Geographical location*	*Type of site* *-Cave (C)* *-Against cliff or rock (R)* *-Freestanding (F)*	*Cultural description*	*Conservation state*	*Excavated*
Desolation island						
Playa Cora D (CoD)	62°° 27′ 36.34″ S 60° 20′ 02.87″ W	In Cora Cove.	F	Rubén Stehberg excavated a fireplace with cultural material at the head of a cove near a freshwater lake identified by Fildes (1821). Cora A has an outline of structural remains, but no structures were defined; Cora C presented electrical insulators, presumably related to the 1935 *Discovery II* expedition visit. Cora D, on the extreme west of Cora Beach, is a very damaged structure with a large fireplace, dump and cultural remains. It claims four indigenous stone artefacts. (Stehberg 2003: 106–129).	Very disturbed sites.	1995
Greenwich island						

Spark Point Greenwich Island	62° 26′ 49.98″ S 59° 43′ 42.6″ W	N-E end of Greenwich Island, on English Strait. Also known as Port Williams.	F	Dry, flat, upper beach terrace that is possibly a seal skin drying floor with a square stone structure, 1.5 × 1.5 m, that is possibly for stacking dried skins.	Inspected in 2007 by Stehberg, Zarankin and Pearson.
Yankee Harbour elephant seal try-works, Greenwich Island	62° 32′ 01.35″ S 59° 46′ 54.24″ W	On the inner end of sand spit forming Yankee Harbour, McFarlane Strait.	F	Recorded by biologist Martin White, 1965/66. Remnant of elephant sealing try-works, with partial disturbed stone arrangement and burnt blubber and soil. A broken try-pot located nearby.	Stable, but tourist site.
Yankee Harbour large enclosure	62° 31′ 55.80″ S 59° 46′ 30.00″ W	On the upper beach terrace, in Gentoo penguin rookery.	F	Recorded by Ronald Lewis-Smith, 1980 and 2006. (Lewis-Smith 2018: 151–158).	Large stone enclosure, 30 × 35 m, with two stone structures within (2.5 × 2.5 m). Purpose and age unknown.

(*Continued*)

APPENDIX (Continued)

Name	*Latitude/ longitude*	*Geographical location*	*Type of site -Cave (C) -Against cliff or rock (R) -Freestanding (F)*	*Cultural description*	*Conservation state*	*Excavated*
King George island						
• Fildes Peninsula						
Cuatro Pircas, Adelia Cove	62° 09′ 57.73″ S 58° 58′ 12.53″ W	The stone structures are situated on a broad beach (known as Rambo Beach after a Brazilian twentieth-century refuge the remains of which survive), between the steep hills and the sea which is about 80 m distant.	F	Four structures are located in close proximity at this site on Adelia Cove. One structure is rectangular without a doorway and is thought to have served as a store into which seal skins were piled. The other rectangular or semi-circular structures are open on one side and are interpreted as being habitation or kitchen structures. Partial excavations uncovered metal containers, glass and ceramics, wood, nails, leather from shoes, textiles and seal skin. The cultural material dates the site to the nineteenth century. (Stehberg 2003: 75–82)	The site is exposed to wind, snow loads and the activity of seals and human visitors.	1984
Punta Pablo Gran, Fontoura Beach (Beach 12, aka Bahia Lobo Marino)	62° 10′ 29.65″ S 58° 58′ 19.96″ W	Against the side of a rock outcrop.	R	The site consists of a stone wall enclosing a rectangular area about 5 m long and 2.4 m wide. Another possible site, approximately 3 × 2 m in extent, is located in the extreme south end of the rock stack, with a few fragments of material from the nineteenth century. (Stehberg 2003: 139–140)	The interior is filled with sand. It was partially excavated in 2001.	2001

Refugio Ruso, Beach 7a	62° 09′ 02.24″ S 58° 56′ 29.49″ W	Located near the Russian Refuge, this semi-circular dry-stone structure is isolated in the central part of a sandy beach.	F		The collapsed walls of the structure stand to approximately 70 cm high, and within it is a whale vertebra, fragments of glass and metal, and pieces of wood; the glass dating from the nineteenth century. The site was partially excavated in 2001. (Stehberg 2003: 135–136, 140–144)	The site is exposed to wind, snow loads and the activity of seals and natural scientists, there being a Russian refuge hut nearby.	2001
Tómbolo 1, try-works	62° 12′ 40.27″ S 59° 01′ 19.07″ W	Located at the western end of the tombolo linking Flat Top Point to the mainland.	F		Two circular stone fireplaces, being furnaces for try-pots used to boil down elephant seal oil, spaced over a 4 m length, with charred bones and ash within each. Nearby is another fireplace and a possible tent site that may have been a small domestic fireplace servicing the sealers. Identified by Pearson and Stehberg 2005.	The site is exposed to wind, snow loads and the activity of seals.	
Tómbolo 2, stone shelter	62° 12′ 43.58″ S 59° 01′ 00.97″ W	Located against the base of the steep hill slope at the western end of the tombolo.	R		A large, square stone-walled structure. A central entry point survives, and within and surrounding the site is an extensive scatter of glass bottle fragments, ceramics, iron barrel hoops, whale bones and shoe leather. The site is interpreted as a nineteenth-century sealer´s occupation place. Identified by Pearson and Stehberg 2007.	The site is exposed to rocky falls of the hill, wind, snow loads and the activity of seals (unrecognised in 2005 as buried in snow).	

(Continued)

APPENDIX (Continued)

Name	*Latitude/ longitude*	*Geographical location*	*Type of site -Cave (C) -Against cliff or rock (R) -Freestanding (F)*	*Cultural description*	*Conservation state*	*Excavated*
Val Klotz site, Beach 9, North Cove	62° 09′ 14.24″ S 58° 58′ 15.39″ W	A natural rock shelter and a stone structure are located adjacent to a low cliff face at the narrow northern end of the beach, close to the sea.	R	The rock shelter when excavated contained some cultural material (ceramics, glass and wood) in the surface levels of the outward sloping ground. The cultural remains were consistent with occupation in the nineteenth century. The rectangular stone structure, c.1.5 × 1.5 m and 30 cm high walls, was partially excavated, and a timber plank with treenail was located on the floor beneath fallen rubble. (Stehberg 2003: 137–138, 144–147).	The rock shelter floor was covered in ice (2005). Stone structure was stable and protected by the cliff above.	2001
• Point Hennequin						
Point Hennequin		Eastern side of Admiralty Bay.		Recorded by biologist Martin White 1965/66 as: 'Stone wall constructed between two former sea stacks 30 [?] yards from present sea level at southern end of peninsula. Most of structure obscured by scree fall'. Described in Lewis-Smith and Simpson (1987: 67) as 'a three-sided stone wall constructed against the north-east side of an outcrop about 10 above high-water mark'. Recorded by Stehberg in the 1990s.		

• Turret Point							
Turret Point	62° 05′ 9.18″ S 57° 57′ 0.42″ W	Opposite Penguin Island, SE coast of KGI.	R		Marked on early maps, Turret Point site of Bransfield territorial claim 1820 (buried bottle of coins). Shelter site on the western side of point against large outcrop, with whale ribs in front. Recorded by biologist Martin White 1965/66. Photo of the site in Lewis-Smith and Simpson (1987: 68) Appears to have been recorded by Stehberg in the 1990s.	Turret Point inspected 2017, sealing shelter site located.	
Livingston Island							
• Byers Peninsula							
Cerro Negro 1–4 (CN 1–4)	62° 39′ 43.08″ S 61° 00′ 11.82″ W	Stone structures against a rock stack, Southern Beaches.	R		The site consists of four stone-walled structures (CN 1–4) of different dimensions against a single rock stack. This site was excavated by Senatore in 1999. It had been occupied at two different times during the nineteenth century, the only multiple-occupation as yet documented. (Zarankin and Senatore 2007: 68–70; Senatore et al. 2018).	Two huts are relatively well preserved, while the remaining two are affected by the action of elephant seals and fur seals.	1998–99

(*Continued*)

APPENDIX (Continued)

Name	*Latitude/ longitude*	*Geographical location*	*Type of site -Cave (C) -Against cliff or rock (R) -Freestanding (F)*	*Cultural description*	*Conservation state*	*Excavated*
Cutler 1 (CU1)	62° 37′ 38.34″ S 60° 59′ 54.18″ W 20 E 0600940 3054986	Northern Beaches.	R	The structure consists of a hut site with three stone walls against a rock stack. There is a high density of artefacts on the surface (fragments of metal, glass, stoneware, among others). This site was excavated by British naturalists in the 1950–80s period. Recorded by biologist Martin White 1965/66, and Lewis-Smith and Simpson (1987: 61–65), who mention the site containing a whale vertebra engraved 'B.S. Cutler SCHr'. (Zarankin and Senatore 2007: 94–94) Benjamin S. Cutler was master of the schooner *Free Gift*, part of the Stonington sealer fleet in 1820–21 and 22–23 (Stackpole 1955: 80).	The stone walls are well preserved despite high impact of the local fauna on the site. No changes have been observed since 1995.	1959–80s (unsystematically)
Cutler 2 (CU2)		Stone structure against a rock stack. Northern Beaches.	R	Structure consisting of three stone walls (Zarankin and Senatore 2007: 96).	This site was identified in 1995 but not was found in 2007. This might have been caused by the effect of erosion and rock falls from the stack.	

Cutler 1 Try-works	63° 37′ 36.24″ S 60° 59′ 56.4″ W	On beach berm at the western end of the beach 100m NE of Cutler 1 shelter site.	F	Two concreted oil and stone circles c.43 cm internal, 75 cm external diameter, 1.8 m apart, bases for try-pots. Flat area 1.3 m^2, 6 m west of pot sites, possible work area.	Located in 2017 by Mike Pearson, in fair condition as surface concretion.	
Lima-Lima (LL)	62° 36′ 55.62″ S 61° 02′ 13.08″ W	At the base of a small rocky hill, fronting the sea. Northern Beaches.	C	The cave is 3,5 m high, 22 m deep and 6 m wide. In the back of the cave, there are two walls forming an entrance to an enclosed area of 4.2 × 2.1 m. Recorded by biologist Martin White 1965/66. Called Lair Point cave by Lewis-Smith and Simpson (1987: 60). This site was excavated by Zarankin and Senatore in 1995–96 (Zarankin and Senatore 2007: 90–91, 124–129, 174–175, Muñoz 2000) and again in 2018–19 by Andrés Zarankin. Thermo-luminescence dates obtained by Rubén Stehberg (pers. com 2008) gave a date range between 1805 and 1830.	While the stone walls are well preserved to date, between 1995 and 2007, a gradual collapse of the cave walls and ceiling was noted.	1959 (unsystematically) 1995–96, 2019

(Continued)

APPENDIX (Continued)

Name	*Latitude/ longitude*	*Geographical location*	*Type of site -Cave (C) -Against cliff or rock (R) -Freestanding (F)*	*Cultural description*	*Conservation state*	*Excavated*
Pencas 1 (PE1)	62° 36′ 26.1″ S 61° 06′ 5.34″ W	Stone walls against rock stack, 15m away from an elephant seal colony. West of Punta Varadero, Northern Beaches.	R	Recorded by biologist Martin White 1965/66. The structure consists of a three-walled stone structure containing three whale vertebra seats. Excavated by Zarankin in 2011. (Zarankin and Senatore 2007: 87).	The stone walls are completely collapsed due to the elephant seal action.	2011
Pencas 2 (PE2)	62° 36′ 24.84″ S 61° 06′ 14.52″ W	Small rock shelter formed between two large sloping rocks. Northern Beaches.	R	Recorded by biologist Martin White 1965/66. The presence of stones in the entrance suggests that a stone wall might have existed. Zarankin found a bottle inside. (Zarankin and Senatore 2007: 88)	Minimal remains visible.	

Pencas 3 (PE3)	62° 36′ 10.62″ S 61° 06′ 20.34″ W	Placed on an open stone terrace on a small peninsula. East of a large berm-enclosed lake. Northern Beaches.	F		Freestanding hut. Straight stone walls enclose a roughly square space. It contains whale vertebra 'seats'. 6 m × 4 m. Excavated by Zarankin in 2011. (Zarankin and Senatore 2007: 89).	This site is poorly preserved.	2011
Playa de los Ladrones 1 (PLL), (Robbery Beach 1)	62° 37′ 19.02″ S 61° 01′ 56.58″ W	Northern Beaches.	R		The structure facing the beach consists of two stone walls enclosing an area between two rock stacks. There are two whale vertebrae in the interior. (Zarankin and Senatore 2007: 91–92).	No changes have been observed since 1995.	
Playa Sur 1 (PS1)	62° 39′ 40.02″ S 60° 58′ 34.56″ W	Low rock ridge adjacent to a beach lagoon, Southern Beaches.	R		The site consists of a stone wall closing a natural square formation formed by a rock outcrop, and an annex. This site was excavated between 1995 and 1997. The main hut was the living area of nineteenth-century sealers, where many artefacts were found. A hearth was found in the annex. (Zarankin and Senatore 2007: 66–67).	The site is well preserved, but the expansion of a nearby lagoon is a concern as it has already covered part of the annex and it is extending to the main hut. Very exposed to sea-level changes and storm surge. 3D recorded in 2017.	1995–96, 1996–97

(Continued)

APPENDIX (Continued)

Name	*Latitude/ longitude*	*Geographical location*	*Type of site -Cave (C) -Against cliff or rock (R) -Freestanding (F)*	*Cultural description*	*Conservation state*	*Excavated*
Playa Sur 2 (PS2)	62° 39′ 41.52″ S, 61° 04′ 11.34″ W	On back of the front beach berm, 40 m from the shore, half-way between Cerro Negro and Sealers Hill, 2.1 km in straight line from Vietor 1, near a channel draining a lagoon behind the beach.	F	Rounded rectangle of medium-sized rocks and one whale vertebra, 2.3–3 m long and 2 m wide.	Located by Zarankin in 2010 and relocated by Pearson and Soares in 2016.	
Punta Diablo 1 (PD1)	62° 40′ 18.66″ S 61° 10′ 42″ W	Against sloping rock outcrop. Devils Point.	R	Stone wall hut contains whale jaw bone suggesting roof support. Fragments of stoneware, glass and metal artefacts are spread inside and around the site. (Zarankin and Senatore 2007: 81–82).	Intensive erosion has occurred due to the action of elephant seals crossing over the site. 3D recorded in 2017.	

Punta Diablo 2 (PD2)	62° 39′ 49.08″ S 61° 09′ 32.7″ W	Located at the northern edge of a penguin rookery on the slope of a hill sheltered by a vertical stack. Devils Point.	F		Reported in the 1950s as the base of wooden hut on a sealer's cargo sledge. Recorded by biologist Martin White 1965/66 as: 'Small hut 8' × 7' hut constructed on sledge runners and planking; charcoal & clinker from a cast iron stove; oil soaked into floor material. At north end of Gentoo rookery, one mile north of Hell Gates'. In 2007, only the remains of a sledge were located and excavated, with no other evidence of occupation. (Zarankin and Senatore 2007: 84–85; Pearson et al 2008; Stehberg et al 2007; Pearson et al. 2009).	Melt water crosses over the site causing erosion.	2007
Punta Elefante 1 (PE1)	62° 41′ 15.82″ S 60° 52' 1.0″ W	West side of Elephant Point.	R		Small enclosed area beside rock outcrop. (Oliva et al. 2016).	Site in good condition	

(Continued)

APPENDIX (Continued)

Name	*Latitude/ longitude*	*Geographical location*	*Type of site -Cave (C) -Against cliff or rock (R) -Freestanding (F)*	*Cultural description*	*Conservation state*	*Excavated*
Punta Elefante 2 (PE2)	62° 41′ 06.9″ S 60° 51′ 13.8″ W	Against main outcrop eastern side of Elephant Point.	R	Very well-preserved stone shelter, with whale rib roof beams stacked at one end, whale vertebra seats and tables, whale jaw bone, fireplace. (Oliva et al. 2016).	Very good condition, assessed as the best-preserved shelter in the region.	2014
Punta Elefante 3 (PE3)	62° 41′ 04.64″ S 60° 51′ 10.8″ W	Two enclosed areas, against rock, faces eastern side of Elephant Point.	R	Two areas. Not excavated. (Oliva et al. 2016).	Site is in good condition.	
Punta Elefante 4 (PE4)	62° 41′ 03.5″ S 60° 51′ 10.8″ W	Against outcrop, eastern side of Elephant Point.	R	Scattered structural stones beside outcrop. (Oliva et al. 2016).	Site is in poor state of preservation.	
Punta Elefante X1 (PEX1)	62° 41′ 06.26″ S 60° 51′ 11.7″ W	Against outcrop eastern side of Elephant Point.	R	Stone walls against rock. Interpreted by Zarankin as work area. Excavated by Zarankin in 2014. (Oliva et al 2016).	Good condition.	2014

Punta Lair 1 (PL1)	62° 36′ 54.78″ S 61° 02′ 06.3″ W	Built against a rock stack. Northern Beaches.	R		Structure consisting of one stone hut and one annex. (Zarankin and Senatore 2007: 93).	This structure is badly preserved due to the action of fauna, water and wind erosion. No changes have been observed since 1995.	
Punta Varadero (PVA)	62° 36′ 29.8″ S 61° 04′ 51.84″ W	Between a rock stack and 1.5m high rocks. Northern Beaches.	R		The structure consists of three stone walls enclosing an area of 2.40 × 2.10 m. The hut contains four whale ribs suggesting roof supports. The annex behind contains 1 whale rib. Excavated by Zarankin in 2011. (Zarankin and Senatore 2007: 85–86)	The stone walls have collapsed.	2011
Punta Vietor 1 (PV1)	62° 40′ 29.04″ S 61° 05′ 42″ W	Against a very large rock stack. Western end of Southern Beaches.	R		Recorded by biologist Martin White 1965/66. The structure consists of three walls against a rock stack enclosing an area of 2.10 × 1.80 m. It contains several whale vertebra 'seats' and a whale rib used for roofing. (Zarankin and Senatore 2007: 71–72)	From 1995 to 2007 some changes have been observed. Five circular holes were found on the rock face immediately behind the hut, indicating the extraction of geological samples (probably geomagnetism). The rear wall of the hut appears to have been pulled down to form a level surface for the geological drilling. 3D recorded in 2017.	

(Continued)

APPENDIX (Continued)

Name	*Latitude/ longitude*	*Geographical location*	*Type of site -Cave (C) -Against cliff or rock (R) -Freestanding (F)*	*Cultural description*	*Conservation state*	*Excavated*
Punta Vietor 2 (PV2)	62° 40′ 29.46″ S 61° 05′ 48.3″ W	Among rock stacks. Western end of Southern Beaches.	R	Recorded by biologist Martin White 1965/66. The site consists of a large stone structure of 5.40 × 2.40 m with an annex of 2.10 × 1.5 m. Walls to 1.3 m high. (Zarankin and Senatore 2007: 72–73)	From 1995 to 2007 some changes have been observed: Some whale vertebra placed inside the inner area of the hut have been moved. 3D recorded in 2017.	
Punta Vietor 3 (PV3)	62° 40′ 29.28″ S 61° 06′ 2.58″ W	Western end of Southern beaches.	F	This structure consists of a 2.5 m long double line of stones. A whale skull bone is placed at the end and continuing one of the line of stones. 4 m long overall. (Zarankin & Senatore 2007: 74). Interpreted by Pearson (2018) as possible boat campsite.	It has limited visibility since the structure does not rise above the ground. 3D recorded in 2017.	
Punta X 1 (PX1)	62° 40′ 16.62″ S 60° 55′ 44.88″ W	Against a rock stack in beach boulders close to the sea and a lagoon separating it from other Punta X sites. Eastern end of Southern Beaches.	R	The structure consists of three stone walls. High impact of faunal movement and erosion has affected the preservation of the site Excavated by Zarankin in 2012. (Zarankin and Senatore 2007: 59).	The whale bones used as roofing support seen in 1995 inside the hut are now almost buried 7m away. 3D recorded in 2017.	2012

Punta X 2 (PX2)	62° 40′ 15″ S 60° 55′ 27.54″ W	Against a big sea stack. Eastern end of Southern Beaches.	R	Stone-walled hut site contains a whale jaw bone suggesting a roof support. Excavated by Zarankin in 2012. (Zarankin and Senatore 2007: 60–61).	The site is poorly preserved due to the action of elephant seals. One of the stone walls is partially collapsed, the other one is 1.5 m tall and is well preserved. Digitally recorded in 2017.	2012
Punta X 3 (PX3)	62° 40′ 16.32″ S 60° 55′ 25.02″ W	Against a rock stack. Eastern end of Southern beaches.	R	Stone-walled hut. Several artefacts including a paddle are visible on the surface site. Three whale ribs on floor. Excavated by Zarankin in 2012. (Zarankin and Senatore 2007: 62–63).	The preservation of this structure is compromised due to the action of elephant seals eroding the surface and stratigraphy of this site. Some changes have been observed since 1995. 3D recorded in 2017.	2012
Rocas Largas (RL)	62° 40′ 26″ S 61° 08′ 32.4″ W	Between rock stack and parallel rock outcrop.	R	An area of 1.20 × 3m in. There is a high density of artefacts (leather shoes, fragments of wood, glass). (Zarankin and Senatore 2007: 82–83).	The presence of some timbers suggests that this area might have been enclosed by a structure in the past.	

(*Continued*)

APPENDIX (Continued)

Name	*Latitude/ longitude*	*Geographical location*	*Type of site -Cave (C) -Against cliff or rock (R) -Freestanding (F)*	*Cultural description*	*Conservation state*	*Excavated*
Sealer 1 (S1)	62° 40′ 26.94″ S 61° 06′ 47.58″ W	Against a rock stack. Western end of Southern beaches.	R	Recorded by biologist Martin White 1965/66. The structure consists of a stone-walled hut and an annex against a rock stack. Lewis Smith and Simpson report the cave above the site to have a stone wall across the back, but no sign of it under deep ice in 2006. Excavated by Zarankin in 2017. (Zarankin and Senatore 2007: 75–76).	The preservation of this structure is compromised due to the action of elephant seals, which often occupy the main hut, thus eroding the surface and stratigraphy of this site. From 1995 to 2007, some changes have been observed: a whale vertebra is now visible on the surface, and the entrance walls of the main hut have collapsed. 3D recorded in 2017.	2017
Sealer 2 (S2)	62° 40′ 21.42″ S 61° 06′ 52.08″ W	Between two rock stacks.	R	A structure made up of a series of curved walls of stone forming a roughly circular space. (Zarankin and Senatore 2007: 76–77).	No changes have been observed from 1995. 3D recorded in 2017	

Sealer 3 (S3)	62° 40′ 23.08″ S 61° 06′ 12.02″ W	On small 3m high outcrops east of Sealers Hill on the western end of Southern Beaches.	R	Structure consisting of a stone-walled hut and an annex containing a whale jaw bone. (Zarankin and Senatore 2007: 78–79; Zarankin et al 2011: 20–25; Moreno 2000; Villagran and Schaefer 2011).	No apparent changes since 1995 have been observed, except the displacement of three whale vertebra from the main hut that have been found at 10m and 20m from the site.	1995, 2010
Sealer 4 (S4)	62° 40′ 20.4″ S 61° 06′ 17.16″ W	Against an isolated low rock outcrop. Western end of Southern beaches.	R	The structure consists of two stone-walled enclosures, one on each side of the outcrop. There are whale vertebra near the site. Excavated by Zarankin in 2010. (Zarankin and Senatore 2007: 79–80).	No changes have been observed since 1995.	2010
Sealer Cave (Cueva Sealer -CuS)	62° 40′ 27.52″ S 61° 06′ 47.10″ W	Cave at the bottom of a cliff. Western end of Southern beaches.	C	Lewis Smith and Simpson reported in the 1950s a stone wall across the rear of the cave, with timbers and seal bones on the floor. (Lewis-Smith and Simpson 1987: 60; Zarankin and Senatore 2007: 80–81).	None of these elements nor the structure can now be found inside the cave.	

(Continued)

APPENDIX (Continued)

Name	*Latitude/ longitude*	*Geographical location*	*Type of site -Cave (C) -Against cliff or rock (R) -Freestanding (F)*	*Cultural description*	*Conservation state*	*Excavated*
Stackpole 1	62° 39′ 54.42″ S 60° 57′ 10.26″ W	On open beach 60 m from Stackpole Stack, eastern end of Southern Beaches.	F	The structure consists of a low line of stones. The purpose of this structure is unclear due to its poor preservation. Possibly an elephant sealing site? (Zarankin and Senatore 2007: 63–64).	It has limited visibility since the structure does not rise above the ground. Much disturbed by seal movement. 3D recorded 2017.	
Stackpole 2	62° 39′ 55.86″ S 60° 56′ 31.74″ W	On open beach terrace, Eastern end of Southern Beaches.	F	The shape of this structure suggests that it could be a grave. It consists of an association of stones and whale bones. Interpreted by Pearson (2018) as boat camp site. (Zarankin and Senatore 2007: 65–66).	Since 1995 some changes have been observed due to the action of people: two whale vertebrae have been moved and appear as if they have been used as seats. 3D recorded in 2017.	
• Cape Shirreff						
Playa Maderas (Pma)	c.62° 27′ 45.89″ S 60° 47′ 09.27″ W	Northern side of Bahia Mansa, NE side of Cape Shirreff.	F	U-shaped stone structure, remains on open beach. Partially excavated. (Stehberg 2003: 96–100).		1994

Playa Yámana 1 (Pya1)	62° 28′ 03.60″ S 60° 48′ 00.12″ W	At the western end of Playa Yámana, Cape Shirreff.	R	Small site against beach outcrop. On the beach where the Amerindian female cranium was found. Site excavated. (Stehberg 2003: 86–96).		1984
Playa Yámana 2 (Pya2)	62° 28′ 03.60″ S 60° 48′ 00.12″ W	At the western end of Playa Yámana, Cape, west side cape Shirreff.	R	Small cave in a beach outcrop immediately adjacent to Playa Yámana 1. Site excavated. (Stehberg 2003: 86–96; Stehberg and Lucero 1996: 59–82; Cartajena and Labarca 2007).		1984
Punta Yusseff	c.62° 28′ 34.73″ S 60° 46′ 16.01″ W	Northern end of Playa Media Luna, east coast Cape Shirreff.	F	Stone pavement/structure at northern end of Half-Moon Bay, in vicinity of where the remains of the *San Telmo* were found in 1820. (Stehberg 2003: 100–101).		
• Siddon's Point						
El Indio (EI)	Approx. 62° 32′ 40″ S 60° 26′ 10″ W	Siddon's Point (aka Punta Negra), north coast, Livingston Island.	R	Complex of six related stone structures. Excavated by Rubén Stehberg. (Stehberg 2003: 102–106).	Unknown (2018).	1994

(*Continued*)

APPENDIX (Continued)

Name	*Latitude/ longitude*	*Geographical location*	*Type of site -Cave (C) -Against cliff or rock (R) -Freestanding (F)*	*Cultural description*	*Conservation state*	*Excavated*
Rugged island						
Rugged Island 1	62° 38′ 04′17″ S 61° 17′ 25.28″ W	Western end of Rugged Island, on the southern inlet.	C	Occupied cave site at the western end of the island. Contained fireplace, scattered leather and moccasins, and elephant seal lance outside. Dated 1820s. Excavated by Rubén Stehberg in 2005. (Pearson and Stehberg 2006).	Site is in good condition.	2005
Rugged Island 2	62° 37′ 28.05″ S 61° 17′ 28.05″ W	Western end of Rugged Island, on the northern inlet.	R	Stone walls against rock outcrop on bay at NW side of island. Main shelter between rocks and annex below. Dated mid-late nineteenth century. Excavated by Rubén Stehberg in 2005. (Pearson and Stehberg 2006).	Site is in good condition.	2005
Snow island						
Presidents Head (Potential site)	62° 43′ 28.08″ S, 61° 12′ 29.03″ W	In a cave on the west of a small beach at the north-western end of Presidents Head.	C	Cave c.10 m deep, 5 m wide, just above high-watermark, with a probable fireplace with burnt stones at the back of the cave. Easy access to the beach and good emergency shelter.	Inspected by Zarankin and Pearson, Jan 2017.	

INDEX

Note: **Bold** page numbers refer to tables; *italic* page numbers refer to figures.

9780367192389